The Quotable Darwin

The Quotable Darwin

Collected and edited by
Janet Browne

PRINCETON UNIVERSITY PRESS
Princeton and Oxford

Copyright © 2018 by Janet Browne

Princeton University Press is committed to the protection of copyright and the intellectual property our authors entrust to us. Copyright promotes the progress and integrity of knowledge created by humans. Thank you for supporting free speech and the global exchange of ideas by purchasing an authorized edition of this book. If you wish to reproduce or distribute any part of it in any form, please obtain permission.

Requests for permission to reproduce material from this work should be sent to Permissions, Princeton University Press

Published by Princeton University Press,
41 William Street, Princeton, New Jersey 08540
In the United Kingdom: Princeton University Press,
99 Banbury Road, Oxford OX2 6JX

press.princeton.edu

Cover image: Charles Darwin by Lock & Whitfield, sepia carbon cabinet card, 1877. © National Portrait Gallery, London.

All Rights Reserved

First paperback printing, 2025
Paperback ISBN 978-0-691-27092-0
Cloth ISBN 978-0-691-16935-4
ISBN (e-book) 978-1-400-88867-2
Library of Congress Cataloging in Publication
Control Number: 2017003239

British Library Cataloging-in-Publication Data is available

This book has been composed in Wilke and Palatino

CONTENTS

List of Illustrations ix
Preface xi
Acknowledgments xix
Chronology xxiii

PART 1: *Early Life and Voyage of the* Beagle

Foundations	3
The *Beagle* Voyage	11
Geology	24
Slavery	30
Natural History Collecting	33
Indigenous Peoples	39
Galápagos Archipelago	45

PART 2: *Marriage and Scientific Work*

Notes on Species	53
Marriage	59
A Theory by Which to Work	65
Children	73
Pigeons	80
Barnacles	84
Precursors	90
Independent Discoveries	97

Part 3: *Origin of Species*

On the Origin of Species	107
Species	117
Selection	122
Difficulties	127
Design and Free Will	133
Variation and Heredity	138
Origin of Life	144
Survival of the Fittest	147
Responses to *On the Origin of Species*	150
Botany	165

Part 4: *Mankind*

Human Origins	173
Race	181
Sexual Selection	185
Morality	190
Intellect	193
Instincts	198
Expression of the Emotions	202
Human Society	207

Part 5: *On Himself*

Religious Belief	217
Health	226
Politics	232
Science	234
Writing	240

Dogs	244
Anti-vivisection	247
Nature	249
Autobiographical	253

Part 6: *Friends and Family*

Friends and Contemporaries	263
Reflections by His Contemporaries	275
Recollections by His Family	286
Tributes	298
Varia	303
Sources	307
Index	315

ILLUSTRATIONS

Charles Darwin, watercolor sketch by George Richmond, 1840	2
Darwin and his son William, unknown artist, daguerreotype, 1842	52
Charles Darwin, photograph by Maull & Fox, c. 1854	106
Caricature, *The Hornet*, 22 March 1871	172
Charles Darwin, photograph by Julia Margaret Cameron, 1868	216
Charles Darwin, photograph by Elliott & Fry, c. 1881	262

PREFACE

Hardly anyone needs an introduction to Charles Darwin, the great nineteenth-century British naturalist who formulated far-reaching ideas about the way in which living beings evolve by natural selection. He is known worldwide for the proposal that all organisms—ourselves included—originate through processes that are entirely natural, without the intervention of any deity, and the label "Darwinian" is habitually used to describe the theory underlying all modern evolutionary biology. This fame is based on his magnificent book, *On the Origin of Species*, which was first published in 1859 in London, amid a storm of controversy, and still stands as one of the foundational texts of the modern world.

But Darwin did much more than publish *Origin of Species*. Earlier in life—aged only 22 at the outset—he traveled on the *Beagle* surveying voyage to South America during the years 1831–36, both as a naturalist and collector and also as a scientific companion to Captain Robert Fitz-Roy. On his return, he continued to carry out natural history researches and observations, work he would pursue for the rest of his life. He was to become recognized as the author of several important scientific publications and an appealing travel book about his experiences on the voyage, called *Journal of Researches,* modeled on the *Personal Narrative* of Alexander von Humboldt. Also on his return he began privately to speculate on evolutionary ideas. He did not arrive at any concrete theory until he read the book published by Thomas Robert Malthus on human population and natural checks to its continual increase. From this Darwin developed his ideas about the struggle for survival and natural selection. Other thinkers at much the same time were proposing evolutionary ideas too, most notably Alfred Russel Wallace, whose work in this area was unsuspected by Darwin until a dramatic moment in 1858 when Wallace wrote to him enclosing an essay about his ideas. The story of the simultaneous announcement of the theory of evolution by natural selection in July 1858 is enthralling. With this announcement Darwin accelerated his plans to publish and produced his big book, *Origin of Species,* the year after.

The debates that followed pushed Darwin and his book into the limelight. These heated discussions also drew in long-standing speculations about the relations between the natural world and its presumed creator, as well as contemporary philosophical inquiries into species and their origins, alongside increasing religious uncertainty, public criticism of the established social order, and rapid industrial and commercial advances in the newly expanding British empire, all of which were reflected in contemporary literature, poetry, and the arts. In a way, Darwin's book crystallized the wide range of issues already in everybody's minds. Indeed, in later decades the book, and the fundamental reassessments it inspired, came to symbolize an important intellectual revolution, helping to make the world modern.

At the center of the storm, Darwin tried to maintain a quiet life. He spent much his time acquiring additional support for the idea of evolution by natural selection—or as he called it, descent with modification—through experimentation on different groups of animals, birds, and, especially, plants, as well as extensive reading and library research. He wrote and published constantly, his work including a series of outstanding later books and articles that elaborated and extended on the theme of evolution by natural selection. Throughout, he responded to reviews and questions, and generated a very large scientific correspondence, eventually stretching across the globe. Without question, he was one of the major authors and thinkers of the period and the catalyst for many people—from all walks of life—to review and perhaps revise their existing vision of the natural world and the place of humankind within it. By the time of his death, he was celebrated as a hero of science. He was buried in Westminster Abbey in London.

What made this extraordinary man tick? Although his books were embedded in public controversy, and his private scientific thoughts were daring, his personal life was extremely conventional. He lived as an independently wealthy Victorian gentleman in the English countryside with a large family and extensive household. After those adventurous years on the *Beagle* voyage, he adopted an unremarkable daily routine, although plagued by continued ill health. He married his cousin Emma Wedgwood in 1839, and together they had ten children, three of whom died before becoming adults. In a series of recollections that he wrote in old age, Darwin described many of the circumstances of his life with great personal modesty. He made it clear that he hated public appearances and was always relieved to let his friends promote his ideas while he stayed quietly at home. Even so, he constantly wrote letters and can perhaps best be seen as participating in the evolutionary controversies by remote control. Luckily for us, the transformations in thought that were shaped by his views mostly took place through the written word.

This volume of quotations from Darwin's writings digs into the historical records to show the remarkable contrasts of his life and times in his own words and in the words of his friends, contemporaries, and family. In print, Darwin was not much given to aphoristic turns of phrase, and he was cautious in the way he expressed his scientific ideas. There are examples of this caution here. However, his private letters and notebooks reveal his thoughts as bold and incisive. His affection for his friends and family is very evident in his correspondence, and he experienced many of the same upsets, family concerns, joy, and grief that other Victorians shared.

The "Quotable" format pioneered by Princeton University Press allows a fresh sort of insight into the individuals chosen as subjects and reflects modern emphasis in historical circles on revealing people fully in the round. As one of the most famous of scientists, Darwin well deserves a volume like this. It provides ready access to the most important ideas that he proposed, the difficulties and criticisms that he encountered, and insight into his personality and family life, all in his own words or those of contemporaries. Included are his time on the *Beagle* voyage, his pleasure in natural history observation, his interest in the question of design in nature, the exciting years when he first stumbled on the idea of evolution, and the composition and reception of his celebrated *Origin of Species*.

Darwin was also a lively letter writer: a man who enjoyed family life and appreciated his friends. Readers will find, for example, that he liked to play billiards because "it drives the horrid species out of my head." His personality shines out from his words, both private and public. Taken as a whole, this book provides a picture of the man as a deeply thoughtful scientist, a talented writer, a loving father, friend, correspondent, and husband.

Perhaps some readers will find that I have overlooked their favorite remarks by Darwin, for which I apologize. Yet I hope that the extracts given here will lead readers further to explore Darwin's letters, notebooks, and published writings; and that he will come alive for them from the written word, as he has for me.

A Note on the Text

The quotations are arranged in short thematic sections for easy access, and the overall format of the volume is broadly chronological. The whole aims to be something more than a comprehensive assemblage of Darwin's best quotes: instead I hope to provide a structured overview of his achievement, set in the context of his own day, along with the responses of his contemporaries—a small volume that will show the trajectory of his thinking on key topics and his public impact. The final pages give Darwin's own reflections on his character and religious beliefs, and remarks about him made by contemporaries and family members. Throughout, the extracts are given verbatim, in Darwin's own spelling and with his own idiosyncratic punctuation and capitalization, without the use of "*sic*." Very occasionally, I have silently made minor changes to help clarify his meaning. Quotation marks have been regularized according to modern usage. Extracts from letters in the Darwin Correspondence online resource are listed with the abbreviation DCP and the letter number. Dates supplied in square brackets are not actually in the letter but are given as established by the research of the Darwin Correspondence Project. Other sources are given in full at the end of the book.

ACKNOWLEDGMENTS

In recent years the study of nineteenth-century science has been transformed by the online publication of documents relating to Darwin's life and work. The Darwin Correspondence Project is a magnificent scholarly resource that makes available in print and online the entire extant correspondence (more than 15,000 items), and much else besides. As well as opening up Darwin's life and work in a way never previously possible, it illuminates significant social and intellectual transformations of the Victorian period. Many of the quotations given here are drawn from letters in this online database, and I offer unreserved thanks to the Project and my warmest appreciation of the scholarship that has gone into this exceptional collection. Permission to publish has graciously been granted by the Syndics of Cambridge University Press and Mr. William Darwin. I particularly thank James A. Secord, Alison Pearn, and the remarkable editorial team of the Project for their friendship over so many years. Other quotations from letters are drawn from published texts that are listed in the sources.

The Complete Work of Charles Darwin Online is another marvelous scholarly resource that makes available multiple editions of every work that Darwin wrote and most of the books he consulted, as well as a wide variety of commentaries and publications about evolutionary theory. This material is essential for understanding Darwin's worldwide impact and the extensive research on which he based his views. I warmly acknowledge the director John van Wyhe in this huge enterprise and record my use of this outstanding website with gratitude and respect. Extracts from Darwin's publications are reproduced with permission from The Complete Work of Charles Darwin Online, John van Wyhe, editor. The two databases together are the most exciting things to happen to the field for many a year.

I wish also to offer my thanks to Katie Ericksen Baca, a former editorial assistant on the Darwin Correspondence Project, now at Harvard University, Department of the History of Science, and a great help to me during the compilation of this volume; and also to Katelyn Smith during the final stages of this book's preparation. I am lucky to have been able to use the resources of the Harvard Library system, Cambridge University Library, and the Wellcome Library, London, for many years. Images are drawn with grateful thanks from the Wellcome Images collection, from English Heritage, the National Portrait Gallery, and University College London. Lastly, I am exceedingly indebted to my editors at Princeton University Press, Alison Kalett and Lauren Bucca, and my friends, students, and colleagues in the History of Science department, Harvard University.

CHRONOLOGY

1809   Born in Shrewsbury, UK, 12 February.

1817   Death of his mother, 15 July.

1825–27 Attends medical classes at Edinburgh University. Meets the evolutionary naturalist Robert Grant.

1827–31 Attends Christ's College, Cambridge University. Meets John Stevens Henslow, Adam Sedgwick, and other renowned professors.

1831   Receives offer to sail on HMS *Beagle*, through J. S. Henslow's recommendation, 29 August. HMS *Beagle* sails from Plymouth, UK, 27 December.

1832   First landfall in Santiago, Cape Verde Islands, followed by Salvador, Brazil, January.

1833   Makes several inland expeditions in Argentina and Uruguay; finds significant fossils around Montevideo, August and November.

1834	First encounters indigenous Fuegians in Tierra del Fuego, February. HMS *Beagle* rounds Cape Horn, April.
1835	HMS *Beagle* expedition surveys the Chilean coast, experiences major earthquake in February, crosses the Andes in March. Visits the Galápagos Islands, September, and Tahiti, November.
1836	Inland excursion in New South Wales, Australia, January. Expedition visits Cocos and Keeling Islands, April.
1836	HMS *Beagle* returns to England, 2 October.
1837	Starts distributing his specimens. Becomes friends with Charles Lyell. Discusses with John Gould the classification of his bird specimens, including the Galápagos finches, January and March. Moves to 36 Great Marlborough Street, London, March. Opens first notebook on transmutation of species, July.
1838–41	Serves as secretary of the Geological Society of London; presents several papers at the Society on his geological findings during the *Beagle* voyage.

1838–43	Edits and superintends publication in five parts *The Zoology of the Voyage of H.M.S. Beagle*.
1838	Reads Malthus and formulates theory of evolution by natural selection, September and October. Moves to 12 Upper Gower Street, London, December.
1839	Marries his cousin Emma Wedgwood, 29 January. Publishes *Journal of Researches into the Natural History and Geology of the Countries Visited during the Voyage of H.M.S. Beagle round the World.* First child, William Erasmus Darwin, born 29 December.
1841	Anne Elizabeth Darwin born 2 March.
1842	Publishes *The Structure and Distribution of Coral Reefs* and writes short sketch of species theory while visiting the Wedgwood family estate Maer, June. Moves to Down House, Kent, September. Mary Eleanor Darwin born 23 September, who dies only a few days later.
1843	Henrietta Emma Darwin born, 25 September.
1844	Completes 230-page essay on species, 5 July. Meets Joseph Dalton Hooker,

who becomes a lifelong friend. George Howard Darwin born 9 July. Robert Chambers's *Vestiges of the Natural History of Creation* published anonymously; Darwin publishes *Geological Observations on the Volcanic Islands Visited during the Voyage of H.M.S. Beagle*.

1845   Publishes second edition of *Journal of Researches*.

1846   Begins study of barnacles; publishes *Geological observations on South America*. Rents land from Sir John Lubbock for the "sandwalk" at Down House.

1847   Elizabeth Darwin born 8 July.

1848   Francis Darwin born 16 August.

1850   Leonard Darwin born 15 January.

1851–54   Publishes two volumes on *Living Cirripedia* and two volumes on *Fossil Cirripedia*.

1851   Death of daughter Anne, aged 10 years, 23 April; Horace Darwin born 13 May.

1854   "Began sorting notes for Species theory."

1855   Begins corresponding with Asa Gray, April.

1856	"Began by Lyell's advice writing species sketch," 14 May. Tenth child, Charles Waring Darwin, born 6 December. Begins to extend Down House to accommodate the growing family, completed 1858. Gets to know Thomas Henry Huxley well.
1858	Receives letter from Alfred Russel Wallace describing Wallace's theory of evolution, 18 June. Darwin "never saw a more striking coincidence." Death of baby Charles from scarlet fever, 28 June. Papers by Darwin and Wallace on the theory of evolution by natural selection read in absentia at the Linnean Society of London, 1 July. Begins writing an "Abstract" that became *On the Origin of Species*, 20 July.
1859	Publishes *On the Origin of Species by Means of Natural Selection, or the Preservation of Favoured Races in the Struggle for Life*, 24 November.
1860	First US edition of *On the Origin of Species* published, New York, January. Controversy at the British Association for the Advancement of Science, Oxford, 30 June. Darwin does not attend.

1861	Publishes third edition of the *Origin of Species*, revised and augmented, including a brief historical sketch of other evolutionary thinkers.
1862	Publishes *On the Various Contrivances by which British and Foreign Orchids are Fertilised by Insects*. Builds a hothouse at Down House for botanical experiments.
1864	Awarded the Copley Medal, the Royal Society of London's highest honor, November.
1865	Publishes *The Movements and Habits of Climbing Plants*.
1868	Publishes *The Variation of Animals and Plants under Domestication*.
1871	Publishes *The Descent of Man, and Selection in Relation to Sex*.
1872	Publishes *The Expression of the Emotions in Man and Animals*. More extensions to Down House, completed 1877.
1875	Publishes *Insectivorous Plants*.
1876	Publishes *The Effects of Cross and Self Fertilisation in the Vegetable Kingdom*. First grandchild, Bernard Darwin, born 7 September, but his mother dies

in childbirth. Bernard and Francis Darwin come to live at Down House. Writes *Recollections of the development of my mind and character*, May to August.

1877 Publishes "A Biographical Sketch of an Infant" and *The Different Forms of Flowers on Plants of the Same Species*.

1879 Publishes a translation of an essay on his grandfather, Dr. Erasmus Darwin, and adds a biographical preface (E. Krause, *Erasmus Darwin*, translated from the German by W. S. Dallas).

1881 Publishes *The Formation of Vegetable Mould Through the Action of Worms, with Observations on Their Habits*.

1882 Dies at Down House, 19 April; buried in Westminster Abbey, London, 26 April.

PART 1

Early Life and the Voyage of the *Beagle*

Charles Darwin watercolor sketch by George Richmond, 1840. Reproduced with permission from Historic England Picture Library. © Historic England Archive.

Foundations

Nothing could have been worse for the development of my mind than Dr. Butler's school [in Shrewsbury], as it was strictly classical, nothing else being taught except a little ancient geography and history. The school as a means of education to me was simply a blank.

Autobiography, 27

Looking back as well as I can at my character during my school life, the only qualities which at this period promised well for the future, were, that I had strong and diversified tastes, much zeal for whatever interested me, and a keen pleasure in understanding any complex subject or thing. I was taught Euclid by a private tutor, and I distinctly remember the intense satisfaction which the clear geometrical proofs gave me.

Autobiography, 43

Towards the close of my school life, my brother worked hard at chemistry and made a fair laboratory with proper apparatus in the tool-house in the garden, and I was allowed to aid him as a servant in most of his experiments. He made all the gases and many compounds, and I read with care several books on chemistry, such as Henry and Parkes' *Chemical Catechism*. The subject interested me greatly, and we often used to go on working till rather late at night. This was the best part of my education at school, for it showed me practically the meaning of experimental science. The fact that we worked at chemistry somehow got known at school, and as it was an unprecedented fact, I was nick-named "Gas."

Autobiography, 45–46

The instruction at Edinburgh [University] was altogether by Lectures, and these were intolerably dull, with the exception of those on chemistry by [T. C.] Hope; but to my mind there are no advantages and many disadvantages in lectures compared with reading. Dr. Duncan's lectures on Materia Medica at 8 o'clock on a winter's morning are something fearful to remember.

Autobiography, 46–47

During my second year in Edinburgh I attended [Robert] Jameson's lectures on Geology and Zoology, but they were incredibly dull. The sole effect they produced on me was the determination never as long as I lived to read a book on Geology or in any way to study the science.

Autobiography, 52

A negro lived in Edinburgh, who had travelled with [Charles] Waterton and gained his livelihood by stuffing birds, which he did excellently; he gave me lessons for payment, and I used often to sit with him, for he was a very pleasant and intelligent man.

Autobiography, 51

I also attended on two occasions the operating theatre in the hospital at Edinburgh, and saw two very bad operations, one on a child, but I rushed away before they were completed. Nor did I ever attend again, for hardly any inducement would have been strong enough to make me do so; this being long before the blessed days of chloroform. The two cases fairly haunted me for many a long year.

Autobiography, 48

During the three years which I spent at Cambridge [University] my time was wasted, as far as the academical studies were concerned, as completely as at Edinburgh and at school.

Autobiography, 58

From my passion for shooting and for hunting and when this failed, for riding across country, I got into a sporting set [at Cambridge University], including some dissipated low-minded young men. We used often to dine together in the evening, though these dinners often included men of a higher stamp, and we sometimes drank too much, with jolly singing and playing at cards afterwards. I know that I ought to feel ashamed of days and evenings thus spent, but as some of my friends were very pleasant and we were all in the highest spirits, I cannot help looking back to these times with much pleasure.

Autobiography, 60

No pursuit at Cambridge was followed with nearly so much eagerness or gave me so much pleasure as collecting beetles. It was the mere passion for collecting, for I did not dissect them and rarely compared their external characters with published descriptions, but got them named anyhow. I will give a proof of my zeal: one day, on tearing off some old bark, I saw two rare beetles and seized one in each hand; then I saw a third and new kind, which I could not bear to lose, so that I popped the one which I held in my right hand into my mouth. Alas it ejected some intensely acrid fluid, which burnt my tongue so that I was forced to spit the beetle out, which was lost, as well as the third one.

Autobiography, 62

When at Cambridge I used to practise throwing up my gun to my shoulder before a looking-glass to see that I threw it up straight. Another and better plan was to get a friend to wave about a lighted candle, and then to fire at it with a cap on the nipple, and if the aim was accurate the little puff of air would blow out the candle. The explosion of the cap caused a sharp crack, and I was told that the Tutor of the College remarked, "What an extraordinary thing it is, Mr Darwin seems to spend hours in cracking a horse-whip in his room, for I often hear the crack when I pass under his windows."

Autobiography, 44–45

I acquired a strong taste for music, and used very often to time my walks so as to hear on weekdays the anthem in King's College Chapel [Cambridge]. This gave me intense pleasure, so that my backbone would sometimes shiver. . . . I am so utterly destitute of an ear, that I cannot perceive a discord, or keep time and hum a tune correctly; and it is a mystery how I could possibly have derived pleasure from music. My musical friends soon perceived my state, and sometimes amused themselves by making me pass an examination, which consisted in ascertaining how many tunes I could recognise, when they were played rather more quickly or slowly than usual. "God save the King" when thus played was a sore puzzle.

Autobiography, 61–62

In order to pass the B.A. examination, it was, also, necessary to get up Paley's *Evidences of Christianity*, and his *Moral Philosophy*. This was done in a thorough manner, and I am convinced that I could have written out the whole of the *Evidences* with perfect correctness, but not of course in the clear language of Paley. The logic of this book and as I may add of his *Natural Theology* gave me as much delight as did Euclid. The careful study of these works, without attempting to learn any part by rote, was the only part of the academical course which, as I then felt and as I still believe, was of the least use to me in the education of my mind. I did not at that time trouble myself about Paley's premises; and taking these on trust I was charmed and convinced by the long line of argumentation.

Autobiography, 59

Whilst examining an old gravel-pit near Shrewsbury a labourer told me that he had found in it a large worn tropical Volute shell, such as may be seen on the chimney-pieces of cottages; and as he would not sell the shell I was convinced that he had really found it in the pit. I told [Professor Adam] Sedgwick of the fact, and he at once said (no doubt truly) that it must have been thrown away by someone into the pit; but then added, if really embedded there it would be the greatest misfortune to geology, as it would overthrow all that we know about the superficial deposits of the midland counties. These gravel-beds belonged in fact to the glacial period, and in after years I found in them broken arctic shells. But I was then utterly astonished at Sedgwick not being delighted at so wonderful a fact as a tropical shell being found near the surface in the middle of England. Nothing before had ever made me thoroughly realize though I had read various scientific books that science consists in grouping facts so that general laws or conclusions may be drawn from them.

Autobiography, 69–70

Before long I became well acquainted with [Professor John Stevens] Henslow, and during the latter half of my time at Cambridge took long walks with him on most days; so that I was called by some of the dons "the man who walks with Henslow".

Autobiography, 64

During my last year at Cambridge I read with care and profound interest [Alexander von] Humboldt's *Personal Narrative*. This work and Sir J. Herschel's *Introduction to the Study of Natural Philosophy* stirred up in me a burning zeal to add even the most humble contribution to the noble structure of Natural Science. No one or a dozen other books influenced me nearly so much as these two.

Autobiography, 67–68

Considering how fiercely I have been attacked by the orthodox it seems ludicrous that I once intended to be a clergyman. Nor was this intention and my father's wish ever formally given up, but died a natural death when on leaving Cambridge I joined the *Beagle* as Naturalist.

Autobiography, 57

The *Beagle* Voyage

I have been asked by [George] Peacock who will read & forward this to you from London to recommend him a naturalist as companion to Capt Fitzroy employed by Government to survey the S. extremity of America. I have stated that I consider you to be the best qualified person I know of who is likely to undertake such a situation—I state this not on the supposition of yr. being a *finished* Naturalist, but as amply qualified for collecting, observing, & noting any thing worthy to be noted in Natural History. . . . Capt. F. wants a man (I understand) more as a companion than a mere collector & would not take anyone however good a Naturalist who was not recommended to him likewise as a *gentleman*.

> J. S. Henslow to Darwin,
> 24 August 1831, DCP 105

My dear Father. . . . I have given Uncle Jos [Josiah Wedgwood II], what I fervently trust is an accurate & full list of your objections, & he is kind enough to give his opinion on all.—The list & his answers will be enclosed.—

(1) Disreputable to my character as a Clergyman hereafter
(2) A wild scheme
(3) That they must have offered to many others before me, the place of Naturalist
(4) And from its not being accepted there must be some serious objection to the vessel or expedition
(5) That I should never settle down to a steady life hereafter
(6) That my accommodations would be most uncomfortable
(7). That you should consider it as again changing my profession
(8) That it would be a useless undertaking

Darwin to R. W. Darwin,
31 August [1831], DCP 110

Gloria in excelsis is the most moderate beginning I can think of.—Things are more prosperous than I should have thought possible.—Cap. Fitzroy is every thing that is delightful, if I was to praise half so much as I feel inclined, you would say it was absurd, only once seeing him.

Darwin to J. S. Henslow,
[5 September 1831], DCP 118

Afterwards on becoming very intimate with Fitz-Roy, I heard that I had run a very narrow risk of being rejected, on account of the shape of my nose! He was an ardent disciple of Lavater, and was convinced that he could judge a man's character by the outline of his features; and he doubted whether anyone with my nose could possess sufficient energy and determination for the voyage. But I think he was afterwards well-satisfied that my nose had spoken falsely.

Autobiography, 72

I do assure you I have been as economical as I possibly could, but my luggage is frightfully bullky—I look forward with consternation to seeing Mr. Wickham [First Lieutenant of the *Beagle*],—if he grumbled merely at the number of my natural cubic inches, what he will do now I cannot imagine.

Darwin to Robert FitzRoy,
[4 or 11 October 1831], DCP 139

My dear Father

I have a long letter, all ready written, but the conveyance by which I send this is so uncertain.—that I will not hazard it, but rather wait for the chance of meeting a homeward bound vessel.—Indeed I only take this opportunity as perhaps you might be anxious, not having sooner heard from me. . . . Natural History goes on excellently & I am incessantly occupied by new & most interesting animals.

> Darwin to R. W. Darwin,
> 10 February 1832, DCP 159

I find to my great surprise that a ship is singularly comfortable for all sorts of work.—Everything is so close at hand, & being cramped, make one so methodical, that in the end I have been a gainer.

> Darwin to R. W. Darwin,
> 8 February–1 March [1832], DCP 158

Nobody who has only been to sea for 24 hours has a right to say, that sea-sickness is even uncomfortable.—The real misery only begins when you are so exhausted—that a little exertion makes a feeling of faintness come on. —I found nothing but lying in my hammock did me any good.

> Darwin to R. W. Darwin,
> 8 February–1 March [1832], DCP 158

It then first dawned on me [on the island of Santiago, Cape Verde] that I might perhaps write a book on the geology of the various countries visited, and this made me thrill with delight. That was a memorable hour to me, and how distinctly I can call to mind the low cliff of lava beneath which I rested, with the sun glaring hot, a few strange desert plants growing near, and with living corals in the tidal pools at my feet. Later in the voyage Fitz-Roy asked to read some of my Journal, and declared it would be worth publishing; so here was a second book in prospect!

Autobiography, 81

Delight itself, however, is a weak term to express the feelings of a naturalist who, for the first time, has been wandering by himself in a Brazilian forest. Among the multitude of striking objects, the general luxuriance of the vegetation bears away the victory. The elegance of the grasses, the novelty of the parasitical plants, the beauty of the flowers, the glossy green of the foliage, all tend to this end. A most paradoxical mixture of sound and silence pervades the shady parts of the wood. The noise from the insects is so loud, that it may be heard even in a vessel anchored several hundred yards from the shore; yet within the recesses of the forest a universal silence appears to reign. To a person fond of natural history, such a day as this, brings with it a deeper pleasure than he ever can hope again to experience.

Journal of Researches 1839, 11

Our beards are all sprouting.—my face at presents looks of about the same tint as a half washed chimney sweeper.—With my pistols in my belt & geological hammer in hand, shall I not look like a grand barbarian?

Darwin to S. E. Darwin,
14 July–7 August [1832], DCP 177

Poor dear old England. I hope my wanderings will not unfit me for a quiet life, & that in some future day, I may be fortunate enough to be qualified to become, like you a country Clergyman. And then we will work together at Nat. History, & I will tell such prodigious stories, as no Baron Monchausen ever did before.—But the Captain says if I indulge in such visions, as green fields & nice little wives &c &c, I shall certainly make a bolt.—So that I must remain contented with sandy plains & great Megatheriums:—

Darwin to W. D. Fox,
[12–13] November 1832, DCP 189

There is high enjoyment in the independence of the Gaucho life—to be able at any moment to pull up your horse, and say, "Here we will pass the night." The deathlike stillness of the plain, the dogs keeping watch, the gipsy-group of Gauchos making their beds round the fire, have left in my mind a strongly-marked picture of this first night, which will not soon be forgotten.

Journal of Researches 1839, 81

One day, as I was amusing myself by galloping and whirling the balls round my head, by accident the free one struck a bush; and its revolving motion being thus destroyed, it immediately fell to the ground, and like magic caught one hind leg of my horse; the other ball was then jerked out of my hand, and the horse fairly secured. Luckily he was an old practised animal, and knew what it meant; otherwise he would probably have kicked till he had thrown himself down. The Gauchos roared with laughter; they cried they had seen every sort of animal caught, but had never before seen a man caught by himself.

Journal of Researches 1839, 51

On arriving at a post-house, we were told by the owner that if we had not a regular passport we must pass on, for there were so many robbers he would trust no one. When he read, however, my passport, which began with "El Naturalista Don Carlos, &c." his respect and civility were as unbounded, as his suspicions had been before. What a naturalist may be, neither he nor his countrymen, I suspect, had any idea; but probably my title lost nothing of its value from that cause.

Journal of Researches 1839, 139

The sort of interest I take in this voyage, is so different a feeling to any thing I ever knew before, that, as in this present instance I have made arrangements for starting [on an inland expedition], all the time knowing I have no business to do it.

Darwin to C. S. Darwin,
13 November 1833, DCP 230

I find being sick at stomach inclines one also to be home-sick.

Darwin to C. S. Darwin,
13 October 1834, DCP 259

The papers will have told you about the great Earthquake of the 20th of February.—I suppose it certainly is the worst ever experienced in Chili.—It is no use attempting to describe the ruins—it is the most awful spectacle I ever beheld.—The town of Concepcion is now nothing more than piles & lines of bricks, tiles & timbers—it is absolutely true there is not one *house* left habitable; some little hovels built of sticks & reeds in the outskirts of the town have not been shaken down & these now are hired by the richest people. The force of the shock must have been immense, the ground is traversed by rents, the solid rocks are shivered, solid buttresses 6–10 feet thick are broken into fragments like so much biscuit.

> Darwin to C. S. Darwin,
> 10–13 March 1835, DCP 271

At night I experienced an attack (for it deserves no less a name) of the *Benchuca* (a species of Reduvius) the great black bug of the Pampas. It is most disgusting to feel soft wingless insects, about an inch long, crawling over one's body. Before sucking they are quite thin, but afterwards become round and bloated with blood, and in this state they are easily crushed. They are also found in the northern parts of Chile and in Peru. One which I caught at Iquique was very empty. When placed on the table, and though surrounded by people, if a finger was presented, the bold insect would immediately draw its sucker, make a charge, and if allowed, draw blood.

> *Journal of Researches* 1839, 403–4

In returning in the evening to the boat [in Tahiti], we stopped to witness a very pretty scene; numbers of children were playing on the beach, and had lighted bonfires, which illuminated the placid sea and surrounding trees. Others, in circles, were singing Tahitian verses. We seated ourselves on the sand, and joined their party. The songs were impromptu, and I believe related to our arrival: one little girl sang a line, which the rest took up in parts, forming a very pretty chorus. The whole scene made us unequivocally aware that we were seated on the shores of an island in the South Sea.

Journal of Researches 1839, 483

After having walked under a burning sun, I do not know any thing more delicious than the milk of a young cocoa-nut. Pine-apples are here so abundant, that the people eat them in the same wasteful manner as we might turnips. They are of an excellent flavour, —perhaps even better than those cultivated in England; and this I believe is the highest compliment which can be paid to a fruit, or indeed to any thing else.

Journal of Researches 1839, 485

At Pahia [in New Zealand], it was quite pleasing to behold the English flowers in the platforms before the houses; there were roses of several kinds, honeysuckle, jasmine, stocks, and whole hedges of sweetbriar.

Journal of Researches 1839, 497

This voyage is terribly long.—I do so earnestly desire to return, yet I dare hardly look forward to the future, for I do not know what will become of me.—Your situation is above envy; I do not venture even to frame such happy visions.—To a person fit to take the office, the life of a Clergyman is a type of all that is respectable & happy: & if he is a Naturalist & has the "Diamond Beetle", ave Maria; I do not know what to say.

Darwin to W. D. Fox,
[9–12 August] 1835, DCP 282

There never was a ship so full of home-sick heroes as the Beagle.

Darwin to E. C. Darwin,
14 February 1836, DCP 298

I loathe, I abhor the sea, & all ships which sail on it.

Darwin to S. E. Darwin,
4 August [1836], DCP 306

Towards the close of our voyage I received a letter whilst at Ascension, in which my sisters told me that [Adam] Sedgwick had called on my father and said that I should take a place among the leading scientific men. I could not at the time understand how he could have learnt anything of my proceedings, but I heard (I believe afterwards) that Henslow had read some of the letters which I wrote to him before the Philosophical Soc. of Cambridge and had printed them for private distribution. My collection of fossil bones, which had been sent to Henslow, also excited considerable attention amongst palaeontologists. After reading this letter I clambered over the mountains of Ascension with a bounding step and made the volcanic rocks resound under my geological hammer! All this shows how ambitious I was.

Autobiography, 81–82

The voyage of the *Beagle* has been by far the most important event in my life and has determined my whole career; yet it depended on so small a circumstance as my uncle offering to drive me 30 miles to Shrewsbury, which few uncles would have done, and on such a trifle as the shape of my nose. I have always felt that I owe to the voyage the first real training or education of my mind. I was led to attend closely to several branches of natural history, and thus my powers of observation were improved, though they were already fairly developed.

Autobiography, 76–77

Geology

The science of Geology is enormously indebted to [Charles] Lyell—more so, as I believe, than to any other man who ever lived. When [I was] starting on the voyage of the Beagle, the sagacious [J. S.] Henslow, who, like all other geologists, believed at that time in successive cataclysms, advised me to get and study the first volume of the "Principles," [Charles Lyell, *Principles of Geology*] which had then just been published, but on no account to accept the views therein advocated. How differently would any one now speak of the "Principles"!

Autobiography, 101

The very first place which I examined, namely St. Jago [Santiago] in the Cape Verde islands, showed me clearly the wonderful superiority of Lyell's manner of treating geology, compared with that of any other author, whose works I had with me or ever afterwards read.

Autobiography, 77

But Geology carries the day; it is like the pleasure of gambling, speculating on first arriving what the rocks may be.

> Darwin to W. D. Fox,
> May [1832], DCP168

I found near the Bajada [Baja de Entre Rios, River Parana, Argentina] a large [fossilized] piece, nearly four feet across, of the giant armadillo-like case; also a molar tooth of a mastodon, and fragments of very many bones, the greater number of which were rotten, and as soft as clay. A tooth which I discovered by one point projecting from the side of a bank, interested me much, for I at once perceived that it had belonged to a horse. Feeling much surprise at this, I carefully examined its geological position, and was compelled to come to the conclusion that a horse . . . lived as a contemporary with the various great monsters that formerly inhabited South America.

> *Journal of Researches* 1839, 149

Having heard of some [fossil] giant's bones at a neighbouring farm-house on the Sarandis [near the River Parana], a small stream entering the Rio Negro, I rode there accompanied by my host, and purchased for the value of eighteen pence, the head of an animal equalling in size that of the hippopotamus. Mr. Owen in a paper read before the Geological Society, has called this very extraordinary animal, Toxodon, from the curvature of its teeth.

Journal of Researches 1839, 180

It is impossible to reflect without the deepest astonishment, on the changed state of this continent [South America]. Formerly it must have swarmed with great monsters, like the southern parts of Africa, but now we find only the tapir, guanaco, armadillo, and capybara; mere pigmies compared to the antecedent races. The greater number, if not all, of these extinct quadrupeds lived at a very recent period; and many of them were contemporaries of the existing molluscs. Since their loss, no very great physical changes can have taken place in the nature of the country. What then has exterminated so many living creatures?

Journal of Researches 1839, 210

The pleasure from the scenery [of the Andes], in itself beautiful, was heightened by the many reflections which arose from the mere view of the grand range, with its lesser parallel ones, and of the broad valley of Quillota directly intersecting the latter. Who can avoid admiring the wonderful force which has upheaved these mountains, and even more so the countless ages which it must have required, to have broken through, removed, and levelled whole masses of them?

Journal of Researches 1839, 314

On the night of the 19th the volcano of Osorno was in activity. At midnight the [ship's] sentry observed something like a large star; from which state the bright spot gradually increased in size till about three o'clock, when a very magnificent spectacle was presented. By the aid of a glass, dark objects, in constant succession, were seen, in the midst of a great red glare of light, to be thrown upwards and to fall down again. The light was sufficient to cast on the water a long bright reflection. By the morning the volcano had resumed its tranquility.

Journal of Researches 1839, 356

The day has been memorable in the annals of Valdivia, for the most severe earthquake experienced by the oldest inhabitant. I happened to be on shore, and was lying down in the wood to rest myself. It came on suddenly, and lasted two minutes; but the time appeared much longer. The rocking of the ground was most sensible. . . . It was something like the movement of a vessel in a little cross ripple, or still more like that felt by a person skating over thin ice, which bends under the weight of his body. A bad earthquake at once destroys the oldest associations: the world, the very emblem of all that is solid, has moved beneath our feet like a crust over a fluid;—one second of time has conveyed to the mind a strange idea of insecurity, which hours of reflection would never have created.

Journal of Researches 1839, 368, 369

The most remarkable effect (or perhaps speaking more correctly, cause) of this earthquake was the permanent elevation of the land. Captain FitzRoy having twice visited the island of Santa Maria, for the purpose of examining every circumstance with extreme accuracy, has brought a mass of evidence in proof of such elevation.

Journal of Researches 1839, 379

No other work of mine was begun in so deductive a spirit as this; for the whole theory [of coral reefs] was thought out on the west coast of S. America before I had seen a true coral reef. I had therefore only to verify and extend my views by a careful examination of living reefs. But it should be observed that I had during the two previous years been incessantly attending to the effects on the shores of S. America of the intermittent elevation of the land, together with denudation and the deposition of sediment. This necessarily led me to reflect much on the effects of subsidence, and it was easy to replace in imagination the continued deposition of sediment by the upward growth of coral. To do this was to form my theory of the formation of barrier reefs and atolls.

Autobiography, 98–99

I always feel as if my books came half out of Lyell's brains, & that I never acknowledge this sufficiently—nor do I know how I can without saying so in so many words—for I have always thought that the great merit of the *Principles*, was that it altered the whole tone of one's mind & therefore that, when seeing a thing never seen by Lyell, one yet saw it partially through his eyes.

Darwin to Leonard Horner,
29 August [1844], DCP 771

Slavery

Early in the voyage at Bahia in Brazil he [Robert FitzRoy] defended and praised slavery, which I abominated, and told me that he had just visited a great slave-owner, who had called up many of his slaves and asked them whether they were happy, and whether they wished to be free, and all answered "No." I then asked him, perhaps with a sneer, whether he thought that the answers of slaves in the presence of their master was worth anything. This made him excessively angry, and he said that as I doubted his word, we could not live any longer together. I thought that I should have been compelled to leave the ship; but as soon as the news spread, which it did quickly, as the captain sent for the first lieutenant to assuage his anger by abusing me, I was deeply gratified by receiving an invitation from all the gun-room officers to mess with them. But after a few hours Fitz-Roy showed his usual magnanimity by sending an officer to me with an apology and a request that I would continue to live with him.

Autobiography, 73–74

I may mention one very trifling anecdote, which at the time struck me more forcibly than any story of cruelty. I was crossing a ferry with a negro, who was uncommonly stupid. In endeavouring to make him understand, I talked loud, and made signs, in doing which I passed my hand near his face. He, I suppose, thought I was in a passion, and was going to strike him; for instantly, with a frightened look and half-shut eyes, he dropped his hands. I shall never forget my feelings of surprise, disgust, and shame, at seeing a great powerful man afraid even to ward off a blow, directed, as he thought, at his face. This man had been trained to a degradation lower than the slavery of the most helpless animal.

Journal of Researches 1839, 28

I have watched how steadily the general feeling, as shown at elections, has been rising against Slavery.—What a proud thing for England, if she is the first Europæan nation which utterly abolishes it.—I was told before leaving England, that after living in Slave countries: all my opinions would be altered; the only alteration I am aware of is forming a much higher estimate of the Negros character.—it is impossible to see a negro & not feel kindly towards him; such cheerful, open honest expressions & such fine muscular bodies; I never saw any of the diminutive Portuguese with their murderous countenances, without almost wishing for Brazil to follow the example of Hayti; & considering the enormous healthy looking black population, it will be wonderful if at some future day it does not take place.

Darwin to E. C. Darwin,
22 May [–14 July] 1833, DCP 206

It does ones heart good to hear how things are going on in England [Act of Parliament for the abolition of slavery in the British Empire]. —Hurrah for the honest Whigs.—I trust they will soon attack that monstrous stain on our boasted liberty, Colonial Slavery.—I have seen enough of Slavery & the dispositions of the negros, to be thoroughly disgusted with the lies & nonsense one hears on the subject in England.

Darwin to J. M. Herbert,
2 June 1833, DCP 209

Natural History Collecting

I mentioned [to Francis Beaufort of the Admiralty], that I believed the Surgeon's collection would be at the disposal of Government and this he thought would make it much easier for me to retain the disposal of my collection amongst the different bodies in London.
>Darwin to Robert FitzRoy,
>[19 September 1831], DCP 131

When at the Rio Negro, in Northern Patagonia, I repeatedly heard the Gauchos talking of a very rare bird which they called Avestruz Petise. They described it as being less than the common (which is there abundant), but with a very close general resemblance. . . . When at Port Desire, in Patagonia (lat. 48°), Mr. Martens shot an ostrich; and I looked at it, forgetting at the moment, in the most unaccountable manner, the whole subject of the Petises, and thought it was a two-third grown one of the common sort. The bird was cooked and eaten before my memory returned. Fortunately the head, neck, legs, wings, many of the larger feathers, and a large part of the skin, had been preserved. From these a very nearly perfect specimen has been put together, and is now exhibited in the museum of the Zoological Society.

Journal of Researches 1839, 108–9

I must have one more growl, by ill luck the French government has sent one of its Collectors to the Rio Negro [Alcide D'Orbigny], —where he has been working for the last six month, & is now gone round the Horn.—So that I am very selfishly afraid he will get the cream of all the good things, before me.

Darwin to J. S. Henslow, [26 October–] 24 November 1832, DCP 192

Amongst the Batrachian reptiles, I found only one little toad, which was most singular from its colour. If we imagine, first, that it had been steeped in the blackest ink, and then when dry, allowed to crawl over a board, freshly painted with the brightest vermilion, so as to colour the soles of its feet and parts of its stomach, a good idea of its appearance will be gained. If it is an unnamed species, surely it ought to be called *diabolicus*, for it is a fit toad to preach in the ear of Eve.

Journal of Researches 1839, 114–15

One evening, when we were about ten miles from the Bay of San Blas [south of Bahía Blanca, Argentina], vast numbers of butterflies, in bands or flocks of countless myriads, extended as far as the eye could range. Even by the aid of a glass it was not possible to see a space free from butterflies. The seamen cried out "it was snowing butterflies," and such in fact was the appearance.

Journal of Researches 1839, 185

A fox, of a kind said to be peculiar to the island [S. Pedro, Chonos Archipelago], and very rare in it, and which is an undescribed species, was sitting on the rocks. He was so intently absorbed in watching their manœuvres [the Beagle surveyors], that I was able, by quietly walking up behind, to knock him on the head with my geological hammer. This fox, more curious or more scientific, but less wise, than the generality of his brethren, is now mounted in the museum of the Zoological Society.

Journal of Researches 1839, 341

On several of the patches of perpetual snow, I found the *Protococcus nivalis*, or red snow, so well known from the accounts of Arctic navigators. My attention was called to the circumstance by observing the footsteps of the mules stained a pale red, as if their hoofs had been slightly bloody. I at first thought it was owing to dust blown from the surrounding mountains of red porphyry; for from the magnifying power of the crystals of snow, the groups of these atom-like plants appeared like coarse particles.

Journal of Researches 1839, 394

In the dusk of the evening [in New South Wales, Australia] I took a stroll along a chain of ponds, which in this dry country represented the course of a river, and had the good fortune to see several of the famous Platypus, or Ornithorhyncus paradoxus. They were diving and playing about the surface of the water, but showed so little of their bodies that they might easily have been mistaken for water-rats. . . . the stuffed specimens do not at all give a good idea of the recent appearance of its head and beak; the latter becoming hard and contracted.

Journal of Researches 1839, 526

A little time before this I had been lying on a sunny bank [in New South Wales, Australia], and was reflecting on the strange character of the animals of this country as compared with the rest of the world. An unbeliever in every thing beyond his own reason might exclaim, "Two distinct Creators must have been at work; their object, however, has been the same, and certainly the end in each case is complete." While thus thinking, I observed the hollow conical pitfall of the lion-ant: first a fly fell down the treacherous slope and immediately disappeared; then came a large but unwary ant; its struggles to escape being very violent, those curious little jets of sand, described by [William] Kirby as being flirted by the insects tail, were promptly directed against the expected victim. But the ant enjoyed a better fate than the fly, and escaped the fatal jaws which lay concealed at the base of the conical hollow. There can be no doubt but that this predacious larva belongs to the same genus with the European kind, though to a different species. Now what would the sceptic say to this? Would any two workmen ever have hit upon so beautiful, so simple, and yet so artificial a contrivance? It cannot be thought so: one Hand has surely worked throughout the universe.

Journal of Researches 1839, 526–27

I find that I did not bring home any Tortoises from the Galapagos, as several were brought home by the surgeon and [Robert] FitzRoy. I have a vague remembrance that specimens were given to the Military Institution in Whitehall (where there is a large model of Waterloo) & I daresay Dr. [John Edward] Gray knows whether this keeps any specimens.

> To Albert Gunther, 12 April [1874?],
> quoted in Gunther 1975, 40

Indigenous Peoples

The Beagle channel was first discovered by Cap FitzRoy during the last voyage, so that it is probable the greater part of the Fuegians had never seen Europeans.—Nothing could exceed their astonishment at the apparition of our four boats: fires were lighted on every point to attract our attention & spread the news.—Many of the men ran for some miles along the shore.—I shall never forget how savage & wild one group was.—Four or five men suddenly appeared on a cliff near to us.—they were absolutely naked & with long streaming hair; springing from the ground & waving their arms around their heads, they sent forth most hideous yells. Their appearance was so strange, that it was scarcely like that of earthly inhabitants.

Beagle Diary, 134

We here saw the native Fuegian; an untamed savage is I really think one of the most extraordinary spectacles in the world.—the difference between a domesticated & wild animal is far more strikingly marked in man.—in the naked barbarian, with his body coated with paint, whose very gestures, whether they may be peacible or hostile are unintelligible, with difficulty we see a fellow-creature.—No drawing or description will at all explain the extreme interest which is created by the first sight of savages.

Darwin to C. S. Darwin,
30 March–12 April 1833, DCP 203

Four natives of Terra del Fuego, were carried to England in the Beagle; were placed under the care of a schoolmaster, in whose house they lived, (one excepted) and there learned to speak English, to use common tools, to plant, and to sow. They were taught the simpler religious truths and duties; and the younger two were beginning to make progress in reading and writing when the time arrived for their return to their own country. I landed them among their people, by whom they were well received, but very soon plundered of most of the treasures their numerous friends in England had given to them. No dulness of apprehension was shewn by those natives—quite the reverse.

FitzRoy and Darwin 1836, 222

All the organs of sense are highly perfected;
sailors are well known for their good eyesight,
& yet the Fuegians were as superior as another
almost would be with a glass.—When Jemmy
[Orundellico, of the Yaghan people] quarrelled
with any of the officers, he would say "me see
ship, me no tell".

Beagle Diary, 137

Jemmy Button now perfectly knew the way
& he guided us to a quiet cove where his family used formerly to reside. We were sorry to
find that Jemmy had quite forgotten his language, that is as far as talking, he could however understand a little of what was said. It
was pitiable, but laughable, to hear him talk
to his brother in English & ask him in Spanish
whether he understood it.

Beagle Diary, 137

It was quite melancholy leaving our Fuegians
amongst their barbarous countrymen: there
was one comfort; they appeared to have no
personal fears.—But, in contradiction of what
has often been stated, 3 years has been sufficient to change savages, into, as far as habits
go, complete & voluntary Europeans. . . . I am
afraid whatever other ends their excursion to
England produces, it will not be conducive to

their happiness.—They have far too much sense not to see the vast superiority of civilized over uncivilized habits; & yet I am afraid to the latter they must return.

Beagle Diary, 142–43

We could hardly recognize poor Jemmy; instead of the clean, well-dressed stout lad we left him, we found him a naked thin squalid savage. York & Fuegia [El'leparu and Yokcushla] had moved to their own country some months ago; the former having stolen all Jemmy's clothes: Now he had nothing, excepting a bit of blanket round his waist.—Poor Jemmy was very glad to see us & with his usual good feeling brought several presents (otter skins which are most valuable to themselves) for his old friends.—The Captain offered to take him to England, but this, to our surprise, he at once refused: in the evening his young wife came alongside & showed us the reason: He was quite contented; last year in the height of his indignation, he said "his country people no *sabe* nothing.—damned fools" now they were very good people, with *too* much to eat & all the luxuries of life.

Darwin to E. C. Darwin,
6 April 1834, DCP 242

The Beagle passed a part of last November at Otaheite or Tahiti.... Mr. Darwin and I landed among a mob of amusing, merry souls, most of them women and children. Mr. Wilson, a missionary who came out in the ship Duff more than thirty years ago, was at the landing place, and welcomed us to his house. The free, cheerful manners of the natives, who gathered about the door, and unceremoniously took possession of vacant seats, either on chairs or on the floor, shewed that they were at home with their instructor, and that churlish seclusion, or affected distance, formed no part of his system.

FitzRoy and Darwin 1836, 224–25

It appears to be forgotten by those persons [critics of the missionaries], that human sacrifices,—the bloodiest warfare,—parricide, —and infanticide,—the power of an idolatrous priesthood,—and a system of profligacy unparalleled in the annals of the world,—have been abolished,—and that dishonesty, licentiousness, and intemperance have been greatly reduced, by the introduction of Christianity. In a voyager it is base ingratitude to forget these things. At the point of shipwreck, how earnestly he will hope that the lesson of the missionary has extended to the place on which he expects to be cast away!

> FitzRoy and Darwin 1836, 228

At sunset, a party of a score of the black aborigines [in Australia] passed by, each carrying, in their accustomed manner, a bundle of spears and other weapons. By giving a leading young man a shilling, they were easily detained, and threw their spears for my amusement. They were all partly clothed, and several could speak a little English; their countenances were good humoured and pleasant; and they appeared far from being such utterly degraded beings as they are usually represented. . . . On the whole they appear to me to stand some few degrees higher in the scale of civilization than the Fuegians.

> *Journal of Researches* 1839, 519

Galápagos Archipelago

The natural history of this archipelago is very remarkable: it seems to be a little world within itself; the greater number of its inhabitants, both vegetable and animal, being found nowhere else.

Journal of Researches 1839, 454–55

It was confidently asserted, that the tortoises coming from different islands in the archipelago were slightly different in form; and that in certain islands they attained a larger average size than in others. Mr. Lawson [the English resident governor] maintained that he could at once tell from which island any one was brought. Unfortunately, the specimens which came home in the Beagle were too small to institute any certain comparison.

Journal of Researches 1839, 465

I was always amused, when overtaking one of these great monsters [a tortoise] as it was quietly pacing along, to see how suddenly, the instant I passed, it would draw in its head and legs, and uttering a deep hiss fall to the ground with a heavy sound, as if struck dead. I frequently got on their backs, and then, upon giving a few raps on the hinder part of the shell, they would rise up and walk away;—but I found it very difficult to keep my balance.

Journal of Researches 1839, 465

I must describe more in detail the tameness of the birds. This disposition is common to all the terrestrial species; namely, to the mocking-birds, the finches, sylvicolæ, tyrant-flycatchers, doves, and hawks. There is not one which will not approach sufficiently near to be killed with a switch, and sometimes, as I have myself tried, with a cap or hat. A gun is here almost superfluous; for with the muzzle of one I pushed a hawk off the branch of a tree. One day a mocking-bird alighted on the edge of a pitcher (made of the shell of a tortoise), which I held in my hand whilst lying down. It began very quietly to sip the water, and allowed me to lift it with the vessel from the ground.

Journal of Researches 1839, 475

This [marine] lizard is extremely common on all the islands throughout the Archipelago. It lives exclusively on the rocky sea-beaches, and is never found, at least I never saw one, even ten yards inshore. It is a hideous-looking creature, of a dirty black colour, stupid and sluggish in its movements.

Journal of Researches 1839, 466–67

One day I carried one [a marine iguana] to a deep pool left by the retiring tide, and threw it in several times as far as I was able. It invariably returned in a direct line to the spot where I stood. It swam near the bottom, with a very graceful and rapid movement, and occasionally aided itself over the uneven ground with its feet. . . . I several times caught this same lizard, by driving it down to a point, and though possessed of such perfect powers of diving and swimming, nothing would induce it to enter the water; and as often as I threw it in, it returned in the manner above described. Perhaps this singular piece of apparent stupidity may be accounted for by the circumstance, that this reptile has no enemy whatever on shore, whereas at sea it must often fall a prey to the numerous sharks. Hence, probably urged by a fixed and hereditary instinct that the shore is its place of safety, whatever the emergency may be, it there takes refuge.

Journal of Researches 1839, 468

These [land] lizards, like their brothers the sea-kind, are ugly animals; and from their low facial angle have a singularly stupid appearance.... I watched one for a long time, till half its body was buried; I then walked up and pulled it by the tail; at this it was greatly astonished, and soon shuffled up to see what was the matter; and then stared me in the face, as much as to say, "What made you pull my tail?"

Journal of Researches 1839, 469–70

When I recollect, the fact that from the form of the body, shape of scales & general size, the Spaniards can at once pronounce, from which Island any tortoise may have been brought, when I see these islands in sight of each other, & possessed of but a scanty stock of animals, tenanted by these birds, but slightly differing in structure & filling the same place in Nature, I must suspect they are only varieties. The only fact of a similar kind of which I am aware, is the constant asserted difference—between the wolf-like Fox of East and West Falkland Islds. —If there is the slightest foundation for these remarks the zoology of Archipelagoes—will be well worth examining; for such facts would undermine the stability of Species.

Ornithological Notes, 262

It never occurred to me, that the productions of islands only a few miles apart, and placed under the same physical conditions, would be dissimilar. I therefore did not attempt to make a series of specimens from the separate islands. It is the fate of every voyager, when he has just discovered what object in any place is more particularly worthy of his attention, to be hurried from it.

Journal of Researches 1839, 474

The archipelago is a little world within itself, or rather a satellite attached to America, whence it has derived a few stray colonists, and has received the general character of its indigenous productions. . . . Hence, both in space and time, we seem to be brought somewhat near to that great fact—that mystery of mysteries—the first appearance of new beings on this earth.

Journal of Researches 1845, 377–78

Seeing this gradation and diversity of structure in one small, intimately related group of birds, one might really fancy that from an original paucity of birds in this archipelago, one species had been taken and modified for different ends.

Journal of Researches 1845, 380

PART 2

Marriage and Scientific Work

Darwin and his son William, unknown artist, daguerreotype, 1842. Reproduced with permission from Historic England Picture Library. © Historic England Archive.

Notes on Species

After my return to England it appeared to me that by following the example of [Charles] Lyell in Geology, and by collecting all facts which bore in any way on the variation of animals and plants under domestication and nature, some light might perhaps be thrown on the whole subject. My first notebook was opened in July 1837. I worked on true Baconian principles, and without any theory collected facts on a wholesale scale, more especially with respect to domesticated productions, by printed enquiries, by conversation with skilful breeders and gardeners, and by extensive reading. . . . I soon perceived that selection was the keystone of man's success in making useful races of animals and plants. But how selection could be applied to organisms living in a state of nature remained for some time a mystery to me.

Autobiography, 119–20

I have been attending a very little to species of birds, & the passages of forms do appear frightful—every thing is arbitrary; no two naturalists agree on any fundamental idea that I can see.

> Darwin to Charles Lyell,
> 30 July 1837, DCP 367

I have lately been sadly tempted to be idle, that is as far as pure geology is concerned, by the delightful number of new views, which have been coming in, thickly & steadily, on the classification & affinities & instincts of animals—bearing on the question of species—note book after note book has been filled with facts which begin to group themselves *clearly* under sub-laws.

> Darwin to Charles Lyell,
> [14] September [1838], DCP 428

In July [1837] opened first note-Book on "Transmutation of Species".—Had been greatly struck from about month of previous March on character of S. American fossils—& species on Galapagos Archipelago.—These facts origin (especially latter) of all my views.

> *Darwin's Journal,* 7

It is absurd to talk of one animal being higher than another.

> *Notebook B,* 74

As [John] Gould remarked to me, the "beauty of species is their exactness," but do not known varieties do the same, May you not breed ten thousand greyhounds & will they not be greyhounds?

Notebook B, 171

People often talk of the wonderful event of intellectual Man appearing—the appearance of insects with other senses is more wonderful.

Notebook B, 206

Why is thought, being a secretion of brain, more wonderful than gravity a property of matter? It is our arrogance, it our admiration of ourselves.

Notebook C, 166

Love of the deity effect of organization. Oh you Materialist!

Notebook C, 166

Man in his arrogance thinks himself a great work, worthy the interposition of a deity, more humble & I believe true to consider him created from animals.

Notebook C, 196–97

One may say there is a force like a hundred thousand wedges trying [to] force every kind of adapted structure into the gaps in the oeconomy of nature, or rather forming gaps by thrusting out weaker ones. The final cause of all this wedgings, must be to sort out proper structure & adapt it to change.

Notebook D, 135

He who understands baboon would do more towards metaphysics than Locke.

Notebook M, 84e

Our descent, then, is the origin of our evil passions!!—The Devil under form of Baboon is our grandfather!

Notebook M, 123

Erasmus [Darwin's brother] says in [Plato's] Phaedo that our "necessary ideas" arise from the preexistence of the soul, are not derivable from experience.—read monkeys for preexistence.

Notebook M, 128

October 8th. Jenny [the orang-utan at London zoo] was amusing herself by getting out ears of corn with her teeth from the straw, & just like child not knowing what to do with them, came several times & opened my hand, & put them in—like child.

Notebook N, 13

During the summer of 1839, and, I believe, during the previous summer, I was led to attend to the cross-fertilisation of flowers by the aid of insects, from having come to the conclusion in my speculations on the origin of species, that crossing played an important part in keeping specific forms constant. I attended to the subject more or less during every subsequent summer; and my interest in it was greatly enhanced by having procured and read in November 1841, through the advice of Robert Brown, a copy of C. K. Sprengel's wonderful book, *Das entdeckte Geheimnis der Natur*.

Autobiography, 127

28th [September 1838] Even the energetic language of [Augustin] Decandolle does not convey the warring of the species as inference from Malthus.—increase of brutes must be prevented solely by positive checks, excepting that famine may stop desire.—in nature production does not increase, whilst no check prevail, but the positive check of famine &

consequently death. I do not doubt every one till he thinks deeply has assumed that increase of animals exactly proportionate to the number that can live.

Notebook D, 134e

In October 1838, that is, fifteen months after I had begun my systematic enquiry, I happened to read for amusement [Thomas Robert] Malthus on *Population*, and being well prepared to appreciate the struggle for existence which everywhere goes on from long-continued observation of the habits of animals and plants, it at once struck me that under these circumstances favourable variations would tend to be preserved, and unfavourable ones to be destroyed. The result of this would be the formation of new species. Here, then, I had at last got a theory by which to work.

Autobiography, 120

Marriage

As for a wife, that most interesting specimen in the whole series of vertebrate animals Providence only knows whether I shall ever capture one or be able to feed her if caught. All such considerations are hidden far in futurity, but at the end of a distant view, I sometimes see a cottage & some white object like a petticoat, which always drives granite & trap out of my head in the most unphilosophical manner.

> Darwin to C. T. Whitley,
> [8 May 1838], DCP 411A

This is the question.

Marry

Children—(if it Please God)—Constant companion, (& friend in old age) who will feel interested in one,—object to be beloved & played with,—better than a dog anyhow.... My God, it is intolerable to think of spending ones whole life, like a neuter bee, working,

working, & nothing after all.—No, no won't do.—Imagine living all one's day solitarily in smoky dirty London House.—Only picture to yourself a nice soft wife on a sofa with good fire, & books & music perhaps—Compare this vision with the dingy reality of Grt. Marlbro' St. [the London house where he was living] Marry—Marry—Marry Q.E.D.

Not Marry

Freedom to go where one liked—choice of Society & *little of it*—Conversation of clever men at clubs. Not forced to visit relatives & to bend in every trifle—to have the expense & anxiety of children—perhaps quarelling— Loss of time.—cannot read in the Evenings— fatness & idleness—Anxiety & responsibility—less money for books &c—if many children forced to gain one's bread. . . . Eheu!! I never should know French,—or see the Continent—or go to America, or go up in a Balloon, or take solitary trip in Wales—poor slave.—you will be worse than a negro—And then horrid poverty (without one's wife was better than an angel & had money)—Never mind my boy—Cheer up—One cannot live this solitary life, with groggy old age, friendless & cold, & childless staring one in ones face, already beginning to wrinkle.—Never mind, trust to chance—keep a sharp look out—There is many a happy slave—

> [July 1838], *Correspondence*,
> vol. 2, 444–45

November 11th. [1838] Sunday. The day of days!

> *Darwin's Journal*, 8

My own dear Emma [his cousin Emma Wedgwood], I kiss the hands with all humbleness and gratitude, which have so filled up for me the cup of happiness—It is my most earnest wish I may make myself worthy of you.

> *Emma Darwin* 1904, vol. 1, 417

My reason tells me that honest & conscientious [religious] doubts cannot be a sin, but I feel it would be a painful void between us. I thank you from my heart for your openness with me & I should dread the feeling that you were concealing your opinions from the fear of giving me pain.

> Emma Wedgwood to Darwin,
> [21–22 November 1838], DCP 441

I believe from your account of your own mind that you will only consider me as a specimen of the genus (I don't know what simia I believe).

You will be forming theories about me & if I am cross or out of temper you will only consider "What does that prove". Which will be a very grand & philosophical way of considering it.

Emma Wedgwood to Darwin,
[23 January 1839], DCP 492

The state of mind that I wish to preserve with respect to you, is to feel that while you are acting conscientiously & sincerely wishing & trying to learn the truth, you cannot be wrong. . . . It seems to me also that the line of your pursuits may have led you to view chiefly the difficulties on one side, & that you have not had time to consider & study the chain of difficulties on the other, but I believe you do not consider your opinions as formed. May not the habit in scientific pursuits of believing nothing till it is proved, influence your mind too much in other things which cannot be proved in the same way, & which if true are likely to be above our comprehension.

Emma Darwin to Darwin,
[c. February 1839]

When I am dead know that many times, I have kissed & cryed over this. C.D.

Note by Darwin, quoted in
Barlow 1958, 236–37

Mem: her beautiful letter to me, safely preserved, shortly after our marriage.

Autobiography, 97

I daresay not a word of this note is really mine; it is all hereditary, except my love for you, which I shd think could not be so, but who knows?

Darwin to Emma Darwin,
[20–21 May 1848], DCP 1176

Our dear old mother [Emma Darwin], who, as you know well, is as good as twice refined gold. Keep her as an example before your eyes, & then [Richard] Litchfield will in future years worship & not only love you, as I worship our dear old mother.

Darwin to his daughter Henrietta on
her marriage to Richard Litchfield,
4 September [1871], DCP 7922

She has been my greatest blessing, and I can declare that in my whole life I have never heard her utter one word which I had rather have been unsaid. She has never failed in the kindest sympathy towards me, and has borne with the utmost patience my frequent complaints from ill-health and discomfort. I do not believe she has ever missed an opportunity of doing a kind action to anyone near her. I marvel at my good fortune that she, so infinitely my superior in every single moral quality, consented to be my wife. She has been my wise adviser and cheerful comforter throughout life, which without her would have been during a very long period a miserable one from ill-health. She has earned the love and admiration of every soul near her.

Autobiography, 96–97

A Theory by Which to Work

At last gleams of light have come, & I am almost convinced (quite contrary to opinion I started with) that species are not (it is like confessing a murder) immutable. . . . I think I have found out (here's presumption) the simple way by which species become exquisitely adapted to various ends.

Darwin to J. D. Hooker,
[11 January 1844], DCP 729

In many genera of insects, and shells, and plants, it seems almost hopeless to establish which are which. In the higher classes there are less doubts; though we find considerable difficulty in ascertaining what deserve to be called species amongst foxes and wolves, and in some birds, for instance in the case of the white barn-owl. When specimens are brought from different parts of the world, how often do naturalists dispute this same question, as I found with respect to the birds brought from the Galapagos islands.

Essay 1844, 82

Let us now suppose a Being with penetration sufficient to perceive differences in the outer and innermost organization quite imperceptible to man, and with forethought extending over future centuries to watch with unerring care and select for any object the offspring of an organism produced under the foregoing circumstances; I can see no conceivable reason why he could not form a new race (or several were he to separate the stock of the original organism and work on several islands) adapted to new ends.

Essay 1844, 85

De Candolle, in an eloquent passage, has declared that all nature is at war, one organism with another, or with external nature. Seeing the contented face of nature, this may at first be well doubted; but reflection will inevitably prove it is too true. The war, however, is not constant, but only recurrent in a slight degree at short periods and more severely at occasional more distant periods; and hence its effects are easily overlooked. It is the doctrine of Malthus applied in most cases with ten-fold force. . . . Lighten any check in the smallest degree, and the geometrical power of increase in every organism will instantly increase the average numbers of the favoured species. Nature may be compared to a surface, on which rest ten thousand sharp wedges touching each other and driven inwards by incessant blows.

Essay 87–88, 89–90

Now can it be doubted from the struggle each individual (or its parents) has to obtain subsistence that any minute variation in structure, habits, or instincts, adapting that individual better to the new conditions, would tell upon its vigour and health? In the struggle it would have a better *chance* of surviving, and those of its offspring which inherited the variation, let it be ever so slight, would have a better *chance* to survive.

Essay 1844, 91

As long as species were thought to be divided and defined by an impassable barrier of *sterility*, whilst we were ignorant of geology, and imagined that the *world was of short duration*, and the number of its past inhabitants few, we were justified in assuming individual creations, or in saying with [William] Whewell that the beginnings of all things are hidden from man.

Essay 1844, 248

My dear Emma,
I have just finished my sketch of my species theory. If, as I believe that my theory is true & if it be accepted even by one competent judge, it will be a considerable step in science. I therefore write this, in case of my sudden death, as my most solemn & last request, which I am sure you will consider the same as if legally entered in my will, that you will devote £400 to its publication.

>Darwin to Emma Darwin,
>5 July 1844, DCP 761

I hate arguments from results, but on my views of descent, really Nat. Hist. becomes a sublimely grand result-giving subject (now you may quiz me for so foolish an escape of mouth).

>Darwin to J. D. Hooker,
>[11–12 July 1845], DCP 889

How painfully (to me) true is your remark that no one has hardly a right to examine the question of species who has not minutely examined many. . . . My only comfort is, (as I mean to attempt the subject) that I have dabbled in several branches of Nat. Hist. & seen good specific men work out my species & know something of geology; (an indispensible union) & though I shall get more kicks than half-pennies, I will, life serving, attempt my work.

Darwin to J. D. Hooker,
[10 September 1845], DCP 915

From September 1854 onwards I devoted all my time to arranging my huge pile of notes, to observing, and experimenting, in relation to the transmutation of species. During the voyage of the *Beagle* I had been deeply impressed by discovering in the Pampean formation great fossil animals covered with armour like that on the existing armadillos; secondly, by the manner in which closely allied animals replace one another in proceeding southwards over the Continent; and thirdly, by the South American character of most of the productions of the Galapagos archipelago, and more especially by the manner in which they differ slightly on each island of the group; none of these islands appearing to be very ancient in a geological sense. It was evident that such facts as these, as well as many others, could be explained on the supposition that species gradually become modified; and the subject haunted me.

Autobiography, 118–19

May 14th [1856] Began by Lyell's advice writing species sketch.

Darwin's Journal, 14

But at that time I overlooked one problem of great importance. . . . This problem is the tendency in organic beings descended from the same stock to diverge in character as they become modified. That they have diverged greatly is obvious from the manner in which species of all kinds can be classed under genera, genera under families, families under sub-orders, and so forth; and I can remember the very spot in the road, whilst in my carriage, when to my joy the solution occurred to me; and this was long after I had come to Down. The solution, as I believe, is that the modified offspring of all dominant and increasing forms tend to become adapted to many and highly diversified places in the economy of nature.

Autobiography, 120–21

What a book a Devil's chaplain might write on the clumsy, wasteful, blundering low & horridly cruel works of nature!

Darwin to J. D. Hooker,
13 July [1856], DCP 1924

The time will come I believe, though I shall not live to see it, when we shall have very fairly true genealogical trees of each great kingdom of nature.

> Darwin to T. H. Huxley,
> 26 September [1857], DCP 2143

I am like Crœsus overwhelmed with my riches in facts. & I mean to make my Book as perfect as ever I can. I shall not go to press at soonest for a couple of years.

> Darwin to W. D. Fox,
> 8 February [1857], DCP 2049

To my mind to say that species were created so & so is no scientific explanation only a reverent way of saying it is so & so.

> Darwin to Asa Gray,
> 20 July [1857], DCP 2125

We have set up a Billiard Table, & I find it does me a deal of good, & drives the horrid species out of my head.

> Darwin to W. D. Fox,
> 24 [March 1859], DCP 2436

It is a mere rag of an hypothesis with as many flaws & holes as sound parts.—My question is whether the rag is worth anything? I think by careful treatment I can carry in it my fruit to market for a short distance over a gentle road; but I fear that you will give the poor rag such a devil of a shake that it will fall all to atoms; & a poor rag is better than nothing to carry one's fruit to market in—So do not be too ferocious.

> Darwin to T. H. Huxley,
> 2 June [1859], DCP 2466

I cannot too strongly express my conviction of the general truth of my doctrines, & God knows I have never shirked a difficulty.

> Darwin to Charles Lyell,
> 20 September [1859], DCP 2492

I fully admit that there are very many difficulties not satisfactorily explained by my theory of descent with modification, but I cannot possibly believe that a false theory would explain so many classes of facts, as I think it certainly does explain.—On these grounds I drop my anchor & believe that the difficulties will slowly disappear.

> Darwin to Asa Gray,
> 11 November [1859], DCP 2520

Children

William: I defy anybody to flatter us on our baby,—for I defy anyone to say anything, in its praise, of which we are not fully conscious. —He is a charming little fellow, & I had not the smallest concepcion there was so much in a five month baby:—You will perceive, by this, that I have a fine degree of paternal fervour.

Darwin to W. D. Fox,
[7 June 1840], DCP 572

William: I saw the first symptom of shyness in my child when nearly two years and three months old: this was shown towards myself, after an absence of ten days from home, chiefly by his eyes being kept slightly averted from mine; but he soon came and sat on my knee and kissed me, and all trace of shyness disappeared.

Darwin 1877, 292

William: It is however extremely difficult to prove that our children instinctively recognise any expression. I attended to this point in my first-born infant, who could not have learnt anything by associating with other children, and I was convinced that he understood a smile and received pleasure from seeing one, answering it by another, at much too early an age to have learnt anything by experience. When this child was about four months old, I made in his presence many odd noises and strange grimaces, and tried to look savage; but the noises, if not too loud, as well as the grimaces, were all taken as good jokes; and I attributed this at the time to their being preceded or accompanied by smiles. When five months old, he seemed to understand a compassionate expression and tone of voice.

Expression, 359

Anne: Another of my infants, a little girl, when exactly a year old, was not nearly so acute, and seemed quite perplexed at the image of a person in a mirror approaching her from behind. The higher apes which I tried with a small looking-glass behaved differently; they placed their hands behind the glass, and in doing so showed their sense, but far from taking pleasure in looking at themselves they got angry and would look no more.

Darwin 1877, 290

CHILDREN 75

Anne: She went to her final sleep most tranquilly, most sweetly at 12 oclock today. Our poor dear dear child has had a very short life but I trust happy, & God only knows what miseries might have been in store for her. She expired without a sigh. How desolate it makes one to think of her frank cordial manners. I am so thankful for the daguerreotype. I cannot remember ever seeing the dear child naughty. God bless her.

Darwin to Emma Darwin,
[23 April 1851], DCP 1412

George: All day long Georgy is drawing ships or soldiers, more especially drummers, whom he will talk about as long as anyone will listen to him.

Darwin to W. E. Darwin,
3 October [1851], DCP 1456

Charles: It was a complete oversight that I did not write to tell you that Emma produced under blessed Chloroform our sixth Boy almost two months ago. I daresay you will think only half-a-dozen Boys a mere joke; but there is a rotundity in the half-dozen which is tremendously serious to me.—Good Heavens to think of all the sendings to School & the Professions afterwards: it is dreadful.

Darwin to W. D. Fox,
8 February [1857], DCP 2049

Charles: It was the most blessed relief to see his poor little innocent face resume its sweet expression in the sleep of death.—Thank God he will never suffer more in this world.

Darwin to J. D. Hooker,
[29 June 1858], DCP 2297

Francis: I am reminded of old days by my third Boy having just begun collecting Beetles, & he caught the other day Brachinus crepitans of immortal Whittlesea-mere memory.—My blood boiled with old ardour, when he caught a Licinus,—a prize unknown to me.

Darwin to W. D. Fox,
13 November [1858], DCP 2360

Francis, Leonard, and Horace Darwin: We three very young collectors have lately taken, in the parish of Down, six miles from Bromley, Kent, the following beetles, which we believe to be rare, namely, *Licinus silphoides*, *Panagus 4-pustulatus* and *Clytus mysticus*. As this parish is only fifteen miles from London, we have thought that you might think it worth while to insert this little notice in the "Intelligencer".

> Darwin to *Entomologist's Weekly Intelligencer*, 25 June 1859, 99

Leonard: I have a Boy with the collecting mania & it has taken the poor form of collecting Postage stamps: he is terribly eager for "Well, Fargo & Co Pony Express 2d & 4d stamp", & in a lesser degree "Blood's 1. Penny Envelope, 1, 3, & 10 cents". If you will make him this present you will give my dear little man as much pleasure, as a new & curious genus gives us old souls.

> Darwin to Asa Gray,
> 10–20 June [1862], DCP 3595

Horace: Horace said to me yesterday, "if everyone would kill adders they would come to sting less". I answered "of course they would, for there would be fewer". He replied indignantly "I did not mean that; but the timid adders which run away would be saved, & in time they would never sting at all." Natural selection of cowards!

> Darwin to John Lubbock,
> 5 September [1862], DCP 3713

Henrietta: From your earliest years you have given me so much pleasure & happiness that you well deserve all the happiness that is possible in return; & I do believe that you are in right way for obtaining it.—I was a favourite of yours before the time when you can remember. How well I can call to mind how proud I was when at Shrewsbury after an absence of a week or fortnight, you would come & sit on my knee, & there you sat for a long time, looking as solemn as a little judge.—Well it is an awful & astounding fact that you are married; & I shall miss you sadly. . . . I shall not look at you as a really married woman, until you are in your own house. It is the furniture which does the job.

> Darwin to Henrietta (Darwin) Litchfield,
> 4 September [1871], DCP 7922

My main objection to them [private schools], as places of education, is the enormous proportion of time spent over classics. I fancy, (though perhaps it is only fancy) that I can perceive the ill & contracting effect on my eldest Boy's mind, in checking interest in anything in which reasoning & observation comes into play.—I shall certainly look out for some school, with more diversified studies for my younger Boys.

Darwin to W. D. Fox,
17 July [1853], DCP 1522

I have indeed been most happy in my family, and I must say to you my children that not one of you has ever given me one minute's anxiety, except on the score of health. There are, I suspect, very few fathers of five sons who could say this with entire truth. When you were very young it was my delight to play with you all, and I think with a sigh that such days can never return. From your earliest days to now that you are grown up, you have all, sons and daughters, ever been most pleasant, sympathetic and affectionate to us and to one another. When all or most of you are at home (as, thank Heavens, happens pretty frequently) no party can be, according to my taste, more agreeable, and I wish for no other society.

Autobiography, 97

Pigeons

I wish you would publish some small fragment of your data, pigeons if you please & so out with the theory & let it take date—& be cited—& understood.

> Charles Lyell to Darwin,
> 1–2 May 1856, DCP 1862

I have now a grand collection of living & dead Pigeons; & I am hand & glove with all sorts of Fanciers, Spitalfield weavers & all sorts of odd specimens of the Human species, who fancy Pigeons.

> Darwin to J. D. Dana,
> 29 September [1856], DCP 1864

I have found my careful work at Pigeons really invaluable, as enlightening me on many points on variation under domestication.

> Darwin to W. D. Fox,
> 3 October [1856], DCP 1867

I sat one evening in a gin-palace in the Borough [Southeast London] amongst a set of pigeon-fanciers,—when it was hinted that

Mr Bult had crossed his Powters with Runts to gain size; & if you had seen the solemn, the mysterious & awful shakes of the head which all the fanciers gave at this scandalous proceeding, you would have recognised how little crossing has had to do with improving breeds, & how dangerous for endless generations the process was.—All this was brought home far more vividly than by pages of mere statements &c.

> Darwin to T. H. Huxley,
> 27 November [1859], DCP 2558

I hope Lady Lyell & yourself will remember whenever you want a little rest & have time how very glad we shd be to see you here, & I will show you my pigeons! which is the greatest treat, in my opinion, which can be offered to human being.

> Darwin to Charles Lyell,
> 4 November [1855], DCP 1772

Sir Charles [Lyell] urged the publication of Mr. D's observations upon pigeons, which he informs me are curious, ingenious, & valuable in the highest degree, accompanied with a brief statement of his general principles. . . . Every body is interested in pigeons.

> Whitwell Elwin to John Murray,
> 3 May 1859, DCP 2457A

Altogether at least a score of pigeons might be chosen, which if shown to an ornithologist, and he were told that they were wild birds, would certainly, I think, be ranked by him as well-defined species. Moreover, I do not believe that any ornithologist would place the English carrier, the short-faced tumbler, the runt, the barb, pouter, and fantail in the same genus; more especially as in each of these breeds several truly-inherited sub-breeds, or species as he might have called them, could be shown him. Great as the differences are between the breeds of pigeons, I am fully convinced that the common opinion of naturalists is correct, namely, that all have descended from the rock-pigeon (*Columba livia*).

Origin 1859, 22–23

I have discussed the probable origin of domestic pigeons at some, yet quite insufficient, length; because when I first kept pigeons and watched the several kinds, knowing well how true they bred, I felt fully as much difficulty in believing that they could ever have descended from a common parent, as any naturalist could in coming to a similar conclusion in regard to the many species of finches, or other large groups of birds, in nature.

Origin 1859, 28

That most skilful breeder, Sir John Sebright, used to say, with respect to pigeons, that "he would produce any given feather in three years, but it would take him six years to obtain head and beak."

Origin 1859, 32

The belief that the chief domestic races are descended from several wild stocks no doubt has arisen from the apparent improbability of such great modifications of structure having been effected since man first domesticated the rock-pigeon. Nor am I surprised at any degree of hesitation in admitting their common origin: formerly, when I went into my aviaries and watched such birds as pouters, carriers, barbs, fantails, and short-faced tumblers, &c., I could not persuade myself that they had all descended from the same wild stock, and that man had consequently in one sense created these remarkable modifications.

Variation 1868, vol. 1, 203–4

Barnacles

When on the coast of Chile, I found a most curious form, which burrowed into the shells of Concholepas, and which differed so much from all other Cirripedes that I had to form a new sub-order for its sole reception. Lately an allied burrowing genus has been found on the shores of Portugal. To understand the structure of my new Cirripede I had to examine and dissect many of the common forms: and this gradually led me on to take up the whole group. I worked steadily on the subject for the next eight years, and ultimately published two thick volumes, describing all the known living species, and two thin quartos on the extinct species. I do not doubt that Sir E. Lytton Bulwer [Edward Bulwer-Lytton] had me in his mind when he introduces in one of his novels a Professor Long, who had written two huge volumes on Limpets.

Autobiography, 117

Are you a good hand at inventing names? I have a quite new & curious genus of Barnacle, which I want to name, & how to invent a name completely puzzles me.

> Darwin to J. D. Hooker,
> [2 October 1846], DCP 1003

There is an extraordinary pleasure in pure observation; not but what I suspect the pleasure in this case is rather derived from comparisons forming in one's mind with allied structures. After having been so many years employed in writing my old geological observations it is delightful to use one's eyes & fingers again.

> Darwin to J. D. Hooker,
> [6 November 1846], DCP 1018

I have lately got a bisexual [two-sexed] cirripede, the male being microscopically small & parasitic within the sack of the female. I tell you this to boast of my species theory, for the nearest & closely allied genus to it is, as usual, hermaphrodite, but I had observed some minute parasites adhering to it, & these parasites, I now can show, are supplemental males, the male organs in the hermaphrodite being unusually small, though perfect & containing zoosperms: so we have almost a polygamous animal, simple females alone being wanting.

> Darwin to J. D. Hooker,
> 10 May 1848, DCP 1174

I never shd have made this out, had not my species theory convinced me that an hermaphrodite species must pass into a bisexual species by insensibly small stages & here we have it, for the male organs in the hermaphrodite are beginning to fail, & independent males already formed. But I can hardly explain what I mean, & you will perhaps wish my Barnacles & Species theory al Diabolo together. But I don't care what you say, my species theory is all gospel.

> Darwin to J. D. Hooker,
> 10 May 1848, DCP 1174

I am now employed on a large volume, describing the anatomy and all the species of barnacles from all over the world. I do not know whether you live near the sea, but if so I should be very glad if you would collect me any that adhere (small and large) to the coast rocks or to shells or to corals thrown up by gales, and send them to me without cleaning out the animals, and taking care of the bases. You will remember that barnacles are conical little shells, with a sort of four-valved lid on the top. There are others with long flexible footstalk, fixed to floating objects, and sometimes cast on shore. I should be very glad of any specimens, but do not give yourself much trouble about them.

> Darwin to Syms Covington
> [Darwin's assistant on the *Beagle*],
> 30 March 1849, DCP 1237

The other day I got the curious case of a unisexual, instead of hermaphrodite, cirripede, in which the female had the common cirripedial character, & in two of the valves of her shell had two little pockets, in *each* of which she kept a little husband. I do not know of any other case where a female invariably has two husbands. . . . Truly the schemes & wonders of nature are illimitable.

> Darwin to Charles Lyell,
> [2 September 1849], DCP 1252

I am particularly obliged to you for pointing out to me your notice on the metamorphosis of the cirripedia in Silliman's Journal, for I sh^d have overlooked it.—You have to a certain extent forestalled me, though we do not take quite the same view on the homologies of the parts.—I have, I think, worked out the anatomy of the larva in considerable detail & I hope correctly.

> Darwin to J. D. Dana,
> 8 October 1849, DCP 1259

You ask what effect studying species has had on my variation theories; I do not think much; I have felt some difficulties more; on the other hand I have been struck. . . . with the variability of every part in some slight degree of every species.

> Darwin to J. D. Hooker,
> 13 June [1850], DCP 1339

I thank you very sincerely for the great trouble you must have taken in collecting so many specimens. I have received a vast number of collections from different places, but never one so rich from one locality. One of the kinds is most curious. It is a new species of a genus of which only one specimen is known to exist in the world, and it is in the British Museum.

> Darwin to Syms Covington,
> 23 November 1850, DCP 1370

Sept. 9th. [1851] Finished packing up all my Cirripedes, preparing fossil balanidae, distributing copies of my work &c. &c. I have yet a few proofs for Fossil Balanidae for Pal. Soc. [the Palaeontological Society] to complete, perhaps a week more work. Began Oct 1, 1846. On Oct 1 it will be 8 years since I began! But then I have lost 1 or 2 years by illness.

> *Darwin's Journal*, 13

My work on the Cirripedia possesses, I think, considerable value, as besides describing several new and remarkable forms, I made out the homologies of the various parts—I discovered the cementing apparatus, though I blundered dreadfully about the cement glands—and lastly I proved the existence in certain genera of minute males complemental to and parasitic on the hermaphrodites. This latter discovery has at last been fully confirmed; though at one time a German writer was pleased to attribute the whole account to my fertile imagination. The Cirripedes form a highly varying and difficult group of species to class; and my work was of considerable use to me, when I had to discuss in the *Origin of Species* the principles of a natural classification. Nevertheless, I doubt whether the work was worth the consumption of so much time.

Autobiography, 117–18

To his children the habit of working at barnacles seemed a commonplace human function, like eating or breathing, and it is reported that one of us being taken into the study of a neighbour, [Sir John Lubbock] and seeing no dissecting table or microscope, asked with justifiable suspicion, "Then where does he do his barnacles?"

F. Darwin 1917, 95

Precursors

I knew him [Robert Grant] well; he was dry and formal in manner, but with much enthusiasm beneath this outer crust. He one day, when we were walking together burst forth in high admiration of Lamarck and his views on evolution. I listened in silent astonishment, and as far as I can judge, without any effect on my mind. I had previously read the *Zoönomia* of my grandfather [Erasmus Darwin], in which similar views are maintained, but without producing any effect on me.

Autobiography, 49

Heaven forfend me from Lamarck nonsense of a "tendency to progression" "adaptations from the slow willing of animals" &c,—but the conclusions I am led to are not widely different from his—though the means of change are wholly so.

Darwin to J. D., Hooker,
[11 January 1844], DCP 729

I was forestalled [by Edward Forbes] in only one important point, which my vanity has always made me regret, namely, the explanation by means of the Glacial period of the presence of the same species of plants and of some few animals on distant mountain summits and in the arctic regions.

Autobiography, 124

I have, also, read the Vestiges [the anonymous evolutionary book by Robert Chambers] but have been somewhat less amused at it, than you appear to have been: the writing & arrangement are certainly admirable, but his geology strikes me as bad, & his zoology far worse.

Darwin to J. D. Hooker,
[7 January 1845], DCP 814

Have you read that strange unphilosophical, but capitally written book, the Vestiges? It has made more talk than any work of late, & has been by some attributed to me—at which I ought to be much flattered & unflattered.

Darwin to W. D. Fox,
[24 April 1845], DCP 859

No educated person, not even the most ignorant, could suppose that I meant to arrogate to myself the origination of the doctrine that species had not been independently created. The only novelty in my work is the attempt to explain how species become modified, & to a certain extent how the theory of descent explains certain large classes of facts; & in these respects I received no assistance from my predecessors.

Darwin to Baden Powell,
18 January [1860], DCP 2655

In last Saturday Gardeners' Chronicle, a Mr Patrick Matthews publishes long extract from his work on "Naval Timber & Arboriculture" published in 1831, in which he briefly but completely anticipates the theory of Nat. Selection.—I have ordered the Book, as some few passages are rather obscure but it, is certainly, I think, a complete but not developed anticipation! Erasmus [Darwin's brother] always said that surely this would be shown to be the case someday. Anyhow one may be excused in not having discovered the fact in a work on "Naval Timber."

Darwin to Charles Lyell,
10 April [1860], DCP 2754

Lamarck was the first man whose conclusions on this subject excited much attention. . . . He upholds the doctrine that all species, including man, are descended from other species. He first did the eminent service of arousing attention to the probability of all change in the organic as well as in the inorganic world being the result of law, and not of miraculous interposition.

Origin 1861, xiii

It is curious how largely my grandfather, Dr. Erasmus Darwin, anticipated the erroneous grounds of opinion, and the views of Lamarck, in his "Zoonomia" (vol. i. p. 500–510), published in 1794. According to Isid. Geoffroy [Isidore Geoffroy Saint Hilaire] there is no doubt that Goethe was an extreme partisan of similar views, as shown in the Introduction to a work written in 1794 and 1795, but not published till long afterwards. It is rather a singular instance of the manner in which similar views arise at about the same period, that Goethe in Germany, Dr. Darwin in England, and Geoffroy Saint Hilaire (as we shall immediately see) in France, came to the same conclusion on the origin of species, in the years 1794–5.

Origin 1861, xiv, note

The "Vestiges of Creation" appeared in 1844. . . . The work, from its powerful and brilliant style, though displaying in the earlier editions little accurate knowledge and a great want of scientific caution, immediately had a very wide circulation. In my opinion it has done excellent service in calling in this country attention to the subject, in removing prejudice, and in thus preparing the ground for the reception of analogous views.

Origin 1861, xv–xvi

Mr. Herbert Spencer, in an Essay (originally published in the "Leader," March 1852, and republished in his "Essays" in 1858), has contrasted the theories of the Creation and the Development of organic beings with remarkable skill and force. He argues from the analogy of domestic productions, from the changes which the embryos of many species undergo, from the difficulty of distinguishing species and varieties, and from the principle of general gradation, that species have been modified; and he attributes the modification to the change of circumstances. The author (1855) has also treated Psychology on the principle of the necessary acquirement of each mental power and capacity by gradation.

Origin 1861, xvii

The "Philosophy of Creation" has been treated in a masterly manner by the Rev. Baden Powell, in his "Essays on the Unity of Worlds," 1855. Nothing can be more striking than the manner in which he shows that the introduction of new species is "a regular, not a casual phenomenon," or, as Sir John Herschel expresses it, "a natural in contra-distinction to a miraculous process."

Origin 1861, xviii

You refer repeatedly to my view as a modification of Lamarck's doctrine of development & progression; if this is your deliberate opinion there is nothing to be said—; but it does not seem so to me; Plato, Buffon, my grandfather before Lamarck & others propounded the obvious view that if species were not created separately, they must have descended from other species: & I can see nothing else in common between the *Origin* & Lamarck. I believe this way of putting the case is very injurious to its acceptance; as it implies necessary progression & closely connects Wallace's & my views with what I consider, after two deliberate readings, as a wretched book; & one from which (I well remember my surprise) I gained nothing.

Darwin to Charles Lyell,
12–13 March [1863], DCP 4038

It has sometimes been said that the success of the Origin proved "that the subject was in the air," or "that men's minds were prepared for it." I do not think that this is strictly true, for I occasionally sounded not a few naturalists, and never happened to come across a single one who seemed to doubt about the permanence of species. Even Lyell and Hooker, though they would listen with interest to me, never seemed to agree. I tried once or twice to explain to able men what I meant by natural selection, but signally failed. What I believe was strictly true is that innumerable well-observed facts were stored in the minds of naturalists, ready to take their proper places as soon as any theory which would receive them was sufficiently explained.

Autobiography, 123–24

Independent Discoveries

But you must not suppose that your paper [On the law which has regulated the introduction of new species, *Annals and Magazine of Natural History* 16 (1855): 184–96] has not been attended to: two very good men, Sir C. Lyell & Mr E. Blyth at Calcutta specially called my attention to it. Though agreeing with you on your conclusion in that paper, I believe I go much further than you; but it is too long a subject to enter on my speculative notions. . . . My work, on which I have now been at work more or less for 20 years, will not fix or settle anything; but I hope it will aid by giving a large collection of facts with one definite end: I get on very slowly, partly from ill-health, partly from being a very slow worker.—I have got about half written; but I do not suppose I shall publish under a couple of years.

Darwin to A. R. Wallace,
22 December 1857, DCP 2192

Some year or so ago, you recommended me to read a paper by [Alfred Russel] Wallace in the Annals, which had interested you & as I was writing to him, I knew this would please him much, so I told him. He has today sent me the enclosed & asked me to forward it to you. It seems to me well worth reading. Your words have come true with a vengeance that I shd be forestalled. You said this when I explained to you here [at Down House] very briefly my views of "Natural Selection" depending on the Struggle for existence.—I never saw a more striking coincidence. If Wallace had my M.S. sketch written out in 1842 he could not have made a better short abstract! Even his terms now stand as Heads of my Chapters.

> Darwin to Charles Lyell,
> 18 [June 1858], DCP 2285

So all my originality, whatever it may amount to, will be smashed.

> Darwin to Charles Lyell,
> 18 [June 1858], DCP 2285

I shd be *extremely* glad now to publish a sketch of my general views in about a dozen pages or so. But I cannot persuade myself that I can do so honourably. . . . I would far rather burn my whole book than that he or any man shd think that I had behaved in a paltry spirit.

> Darwin to Charles Lyell,
> [25 June 1858], DCP 2294

It is miserable in me to care at all about priority.

> Darwin to J. D. Hooker,
> [29 June 1858], DCP 2298

Mr Wallace who is now exploring New Guinea, has sent me an abstract of the same theory, most curiously coincident even in expressions. And he could never have heard a word of my views. He directed me to forward it to [Charles] Lyell.—Lyell who is acquainted with my notions consulted with [Joseph D.] Hooker, (who read a dozen years ago a long sketch of mine written in 1844) urged me with much kindness not to let myself to be quite forestalled & to allow them to publish with Wallace's paper an abstract of mine; & as the only very brief thing which I had written out was a copy of my letter to you, I sent it and, I believe, it has just been read, (though never written, & not fit for such purpose) before the Linnean Socy.

> Darwin to Asa Gray,
> 4 July 1858, DCP 2302

The accompanying papers, which we have the honour of communicating to the Linnean Society, and which all relate to the same subject . . . contain the results of the investigations of two indefatigable naturalists, Mr. Charles Darwin and Mr. Alfred Wallace. These gentlemen having, independently and unknown to one another, conceived the same very ingenious theory to account for the appearance and perpetuation of varieties and of specific forms on our planet, may both fairly claim the merit of being original thinkers in this important line of inquiry; but neither of them having published his views, though Mr. Darwin has for many years past been repeatedly urged by us to do so, and both authors having now unreservedly placed their papers in our hands, we think it would best promote the interests of science that a selection from them should be laid before the Linnean Society.

> Charles Lyell and J. D. Hooker, quoted in Darwin and Wallace 1858, 45

I always thought it very possible that I might be forestalled, but I fancied that I had grand enough soul not to care; but I found myself mistaken & punished; I had, however, quite resigned myself & had written half a letter to Wallace to give up all priority to him & sh^d

certainly not have changed had it not been for Lyell's & yours quite extraordinary kindness. I assure you I feel it, & shall not forget it.

> Darwin to J. D. Hooker,
> 13 [July 1858], DCP 2306

It would have caused me much pain & regret had Mr. Darwin's excess of generosity led him to make public my paper unaccompanied by his own much earlier & I doubt not much more complete views on the same subject, & I must again thank you for the course you have adopted, which while strictly just to both parties, is so favourable to myself.

> Alfred Russel Wallace to J. D. Hooker,
> 6 October 1858, DCP 2337

I enclose letters to you & me from Wallace. I admire extremely the spirit in which they are written. I never felt very sure what he would say. He must be an amiable man. Please return that to me, & Lyell ought to be told how well satisfied he is.—These letters have vividly brought before me how much I owe to your & Lyell's most kind & generous conduct in all this affair.

> Darwin to J. D. Hooker,
> 23 January [1859], DCP 2403

Mr. Wallace, who is now studying the natural history of the Malay archipelago, has arrived at almost exactly the same general conclusions that I have on the origin of species. Last year he sent to me a memoir on this subject, with a request that I would forward it to Sir Charles Lyell, who sent it to the Linnean Society, and it is published in the third volume of the Journal of that Society. Sir C. Lyell and Dr. Hooker, who both knew of my work—the latter having read my sketch of 1844—honoured me by thinking it advisable to publish, with Mr. Wallace's excellent memoir, some brief extracts from my manuscripts.

Origin 1859, 1–2

The extract from my MS. and the letter to Asa Gray had neither been intended for publication, and were badly written. Mr Wallace's essay, on the other hand, was admirably expressed and quite clear. Nevertheless, our joint productions excited very little attention, and the only published notice of them which I can remember was by Professor Haughton of Dublin, whose verdict was that all that was new in them was false, and what was true was old.

Autobiography, 122

> The year which has passed . . . has not, indeed, been marked by any of those striking discoveries which at once revolutionize, so to speak, the department of science on which they bear.
>
> Thomas Bell, presidential address,
> *Proceedings of the Linnean Society of London* 1859, viii

PART 3

Origin of Species

Darwin, photograph by Maull & Fox, c. 1854. Reproduced with permission from National Portrait Gallery, London.

On the Origin of Species

WHEN on board H.M.S. "Beagle," as naturalist, I was much struck with certain facts in the distribution of the inhabitants of South America, and in the geological relations of the present to the past inhabitants of that continent. These facts seemed to me to throw some light on the origin of species—that mystery of mysteries, as it has been called by one of our greatest philosophers.

Origin 1859, 1

The Struggle for Existence amongst all organic beings throughout the world, which inevitably follows from their high geometrical powers of increase, will be treated of. This is the doctrine of Malthus, applied to the whole animal and vegetable kingdoms. As many more individuals of each species are born than can possibly survive; and as, consequently, there is a frequently recurring struggle for existence, it follows that any being, if it vary however slightly in any manner profitable to itself, under the complex and sometimes varying conditions of life, will have a better chance of surviving, and thus be *naturally selected*. From the strong principle of inheritance, any selected variety will tend to propagate its new and modified form.

Origin 1859, 4–5

I am fully convinced that species are not immutable; but that those belonging to what are called the same genera are lineal descendants of some other and generally extinct species, in the same manner as the acknowledged varieties of any one species are the descendants of that species. Furthermore, I am convinced that Natural Selection has been the main but not exclusive means of modification.

Origin 1859, 6

How have all those exquisite adaptations of one part of the organisation to another part, and to the conditions of life, and of one distinct organic being to another being, been perfected? We see these beautiful co-adaptations most plainly in the woodpecker and missletoe; and only a little less plainly in the humblest parasite which clings to the hairs of a quadruped or feathers of a bird; in the structure of the beetle which dives through the water; in the plumed seed which is wafted by the gentlest breeze; in short, we see beautiful adaptations everywhere and in every part of the organic world.

Origin 1859, 60–61

We behold the face of nature bright with gladness, we often see superabundance of food; we do not see, or we forget, that the birds which are idly singing round us mostly live on insects or seeds, and are thus constantly destroying life; or we forget how largely these songsters, or their eggs, or their nestlings, are destroyed by birds and beasts of prey; we do not always bear in mind, that though food may be now superabundant, it is not so at all seasons of each recurring year.

Origin 1859, 62

Lighten any check, mitigate the destruction ever so little, and the number of the species will almost instantaneously increase to any amount. The face of Nature may be compared to a yielding surface, with ten thousand sharp wedges packed close together and driven inwards by incessant blows, sometimes one wedge being struck, and then another with greater force.

Origin 1859, 66–67

From experiments which I have tried, I have found that the visits of bees, if not indispensable, are at least highly beneficial to the fertilisation of our clovers; but humble-bees alone visit the common red clover (Trifolium pratense), as other bees cannot reach the nectar. Hence I have very little doubt, that if the whole genus of humble-bees became extinct or very rare in England, the heartsease and red clover would become very rare, or wholly disappear. The number of humble-bees in any district depends in a great degree on the number of field-mice, which destroy their combs and nests; and Mr. H. Newman, who has long attended to the habits of humble-bees, believes that "more than two thirds of them are thus destroyed all over England." Now the number of mice is largely dependent, as every one knows, on the number of cats; and Mr. Newman says, "Near villages and small towns I have found the nests of humble-bees more numerous than elsewhere, which I attribute to the number of cats that destroy the mice." Hence it is quite credible that the presence of a feline animal in large numbers in a district might determine, through the intervention first of mice and then of bees, the frequency of certain flowers in that district!

Origin 1859, 73–74

Can the principle of selection, which we have seen is so potent in the hands of man, apply in nature? I think we shall see that it can act most effectually.

Origin 1859, 80

In North America the black bear was seen by Hearne swimming for hours with widely open mouth, thus catching, like a whale, insects in the water. Even in so extreme a case as this, if the supply of insects were constant, and if better adapted competitors did not already exist in the country, I can see no difficulty in a race of bears being rendered, by natural selection, more and more aquatic in their structure and habits, with larger and larger mouths, till a creature was produced as monstrous as a whale.

Origin 1859, 184

The affinities of all the beings of the same class have sometimes been represented by a great tree. I believe this simile largely speaks the truth. The green and budding twigs may represent existing species; and those produced during each former year may represent the long succession of extinct species. . . . As buds give rise by growth to fresh buds, and these, if vigorous, branch out and overtop on all sides many a feebler branch, so by generation I believe it has been with the great Tree of Life, which fills with its dead and broken branches the crust of the earth, and covers the surface with its ever branching and beautiful ramifications.

Origin 1859, 129–30

Many naturalists think that something more is meant by the Natural System [of classification]; they believe that it reveals the plan of the Creator; but unless it be specified whether order in time or space, or what else is meant by the plan of the Creator, it seems to me that nothing is thus added to our knowledge.

Origin 1859, 413

We can see why characters derived from the embryo should be of equal importance with those derived from the adult, for our classifications of course include all ages of each species. But it is by no means obvious, on the ordinary view, why the structure of the embryo should be more important for this purpose than that of the adult, which alone plays its full part in the economy of nature.

Origin 1859, 418–19

The embryos, also, of distinct animals within the same class are often strikingly similar: a better proof of this cannot be given, than a circumstance mentioned by [Louis] Agassiz, namely, that having forgotten to ticket the embryo of some vertebrate animal, he cannot now tell whether it be that of a mammal, bird, or reptile.

Origin 1859, 439

Embryology rises greatly in interest, when we thus look at the embryo as a picture, more or less obscured, of the common parent-form of each great class of animals.

Origin 1859, 450

This whole volume is one long argument.

Origin 1859, 459

I by no means expect to convince experienced naturalists whose minds are stocked with a multitude of facts all viewed, during a long course of years, from a point of view directly opposite to mine. It is so easy to hide our ignorance under such expressions as the "plan of creation," "unity of design," &c., and to think that we give an explanation when we only restate a fact.

Origin 1859, 481–82

Authors of the highest eminence seem to be fully satisfied with the view that each species has been independently created. To my mind it accords better with what we know of the laws impressed on matter by the Creator, that the production and extinction of the past and present inhabitants of the world should have been due to secondary causes, like those determining the birth and death of the individual.

Origin 1859, 488

When the views entertained in this volume on the origin of species, or when analogous views are generally admitted, we can dimly foresee that there will be a considerable revolution in natural history.

Origin 1859, 484

When we no longer look at an organic being as a savage looks at a ship, as at something wholly beyond his comprehension; when we regard every production of nature as one which has had a history; when we contemplate every complex structure and instinct as the summing up of many contrivances, each useful to the possessor, nearly in the same way as when we look at any great mechanical invention as the summing up of the labour, the experience, the reason, and even the blunders of numerous workmen; when we thus view each organic being, how far more interesting, I speak from experience, will the study of natural history become!

Origin 1859, 485–86

In the distant future I see open fields for far more important researches. Psychology will be based on a new foundation, that of the necessary acquirement of each mental power and capacity by gradation. Light will be thrown on the origin of man and his history.

Origin 1859, 488

It is interesting to contemplate an entangled bank, clothed with many plants of many kinds, with birds singing on the bushes, with various insects flitting about, and with worms crawling through the damp earth, and to reflect that these elaborately constructed forms, so different from each other, and dependent on each other in so complex a manner, have all been produced by laws acting around us.

Origin 1859, 489

There is grandeur in this view of life, with its several powers, having been originally breathed into a few forms or into one; and that, whilst this planet has gone cycling on according to the fixed law of gravity, from so simple a beginning endless forms most beautiful and most wonderful have been, and are being, evolved.

Origin 1859, 490

I gained much by my delay in publishing from about 1839, when the theory was clearly conceived, to 1859; and I lost nothing by it, for I cared very little whether men attributed most originality to me or Wallace; and his essay no doubt aided in the reception of the theory.

Autobiography, 124

It is no doubt the chief work of my life.

Autobiography, 122

Species

No one definition has as yet satisfied all naturalists; yet every naturalist knows vaguely what he means when he speaks of a species. Generally the term includes the unknown element of a distinct act of creation. The term "variety" is almost equally difficult to define; but here community of descent is almost universally implied, though it can rarely be proved.

Origin 1859, 44

Certainly no clear line of demarcation has as yet been drawn between species and sub-species—that is, the forms which in the opinion of some naturalists come very near to, but do not quite arrive at the rank of species; or, again, between sub-species and well-marked varieties, or between lesser varieties and individual differences. These differences blend into each other in an insensible series; and a series impresses the mind with the idea of an actual passage.

Origin 1859, 50–51

From these remarks it will be seen that I look at the term species, as one arbitrarily given for the sake of convenience to a set of individuals closely resembling each other, and that it does not essentially differ from the term variety, which is given to less distinct and more fluctuating forms. The term variety, again, in comparison with mere individual differences, is also applied arbitrarily, and for mere convenience sake.

Origin 1859, 52

Alphonse De Candolle and others have shown that plants which have very wide ranges generally present varieties; and this might have been expected, as they become exposed to diverse physical conditions, and as they come into competition (which, as we shall hereafter see, is a far more important circumstance) with different sets of organic beings. But my tables further show that, in any limited country, the species which are most common, that is abound most in individuals, and the species which are most widely diffused within their own country (and this is a different consideration from wide range, and to a certain extent from commonness), often give rise to varieties sufficiently well-marked to have been recorded in botanical works. Hence it is the most flourishing, or, as they may be called, the dominant species,—those which range widely over the world, are the most diffused in their own country, and are the most numerous in individuals,—which oftenest produce well-marked varieties, or, as I consider them, incipient species.

Origin 1859, 53–54

I conclude that, although small isolated areas probably have been in some respects highly favourable for the production of new species, yet that the course of modification will generally have been more rapid on large areas; and what is more important, that the new forms produced on large areas, which already have been victorious over many competitors, will be those that will spread most widely, will give rise to most new varieties and species, and will thus play an important part in the changing history of the organic world.

Origin 1859, 106

The more diversified the descendants from any one species become in structure, constitution, and habits, by so much will they be better enabled to seize on many and widely diversified places in the polity of nature, and so be enabled to increase in numbers.

Origin 1859, 112

I am inclined to believe that almost every species (as we see with nearly all our domestic productions) varies sufficiently for natural selection to pick out & accumulate new specific differences, under new organic & inorganic conditions of life; whenever a place is open in the polity of nature. But that looking to *a long lapse of time* & to the whole world or to large parts of the world, I believe only one or a few species of each large genus ultimately becomes victorious & leaves modified descendants.

Darwin to Charles Lyell,
3 October [1860], DCP 2935

There is one point connected with individual differences, which seems to me extremely perplexing: I refer to those genera which have sometimes been called "protean" or "polymorphic," in which the species present an inordinate amount of variation; and hardly two naturalists can agree which forms to rank as species and which as varieties. We may instance Rubus, Rosa, and Hieracium amongst plants, several genera of insects, and several genera of Brachiopod shells. In most polymorphic genera some of the species have fixed and definite characters. Genera which are polymorphic in one country seem to be, with some few exceptions, polymorphic in other countries, and likewise, judging from Brachiopod shells, at former periods of time. These facts seem to be very perplexing, for they seem to show that this kind of variability is independent of the conditions of life.

Origin 1859, 46

Selection

One of the most remarkable features in our domesticated races is that we see in them adaptation, not indeed to the animal's or plant's own good, but to man's use or fancy. . . . The key is man's power of accumulative selection: nature gives successive variations; man adds them up in certain directions useful to him. In this sense he may be said to make for himself useful breeds.

Origin 1859, 29–30

It may be said that natural selection is daily and hourly scrutinising, throughout the world, every variation, even the slightest; rejecting that which is bad, preserving and adding up all that is good; silently and insensibly working, whenever and wherever opportunity offers, at the improvement of each organic being in relation to its organic and inorganic conditions of life. We see nothing of these slow changes in progress, until the hand of time has marked the long lapse of ages, and then so imperfect is our view into long past geological ages, that we only see that the forms of life are now different from what they formerly were.

Origin 1859, 84

Natural selection cannot possibly produce any modification in any one species exclusively for the good of another species. . . . If it could be proved that any part of the structure of any one species had been formed for the exclusive good of another species, it would annihilate my theory, for such could not have been produced through natural selection.

Origin 1859, 200–201

Natural selection tends only to make each organic being as perfect as, or slightly more perfect than, the other inhabitants of the same country with which it has to struggle for existence. And we see that this is the degree of perfection attained under nature. The endemic productions of New Zealand, for instance, are perfect one compared with another; but they are now rapidly yielding before the advancing legions of plants and animals introduced from Europe. Natural selection will not produce absolute perfection, nor do we always meet, as far as we can judge, with this high standard under nature.

Origin 1859, 202

One word more upon the "Deification" of Natural Selection. Attributing so much weight to it, does not exclude still more general laws i.e. the ordering of the whole universe. I have said that nat. selection is to the structure of organised beings, what the human architect is to a building. The very existence of the human architect shows the existence of more general laws; but no one in giving credit for a building to the human architect, thinks it necessary to refer to the laws by which man has appeared.

>Darwin to Charles Lyell,
>17 June [1860], DCP 2833

I do *not* agree with your remark that I make N. Selection do too much work.—You will perhaps reply, that every man rides his Hobbyhorse to death; & that I am in this galloping state.

>Darwin to Charles Lyell,
>3 October [1860], DCP 2937

When this or that part has been spoken of as contrived for some special purpose, it must not be supposed that it was originally always formed for this sole purpose. The regular course of events seems to be, that a part which originally served for one purpose, by slow changes becomes adapted for widely different purposes.... On the same principle, if a man were to make a machine for some special purpose, but were to use old wheels, springs, and pulleys, only slightly altered, the whole machine, with all its parts, might be said to be specially contrived for that purpose. Thus throughout nature almost every part of each living being has probably served, in a slightly modified condition, for diverse purposes, and has acted in the living machinery of many ancient and distinct specific forms.

Orchids, 346, 348

It is evidently also necessary not to personify "nature" too much,—though I am very apt to do it myself,—since people will not understand that all such phrases are metaphors.

A.R. Wallace to Darwin,
2 July 1866, DCP 5145

For brevity sake I sometimes speak of natural selection as an intelligent power;—in the same way as astronomers speak of the attraction of gravity as ruling the movements of the planets, or as agriculturists speak of man making domestic races by his power of selection. In the one case, as in the other, selection does nothing without variability, and this depends in some manner on the action of the surrounding circumstances on the organism. I have, also, often personified the word Nature; for I have found it difficult to avoid this ambiguity; but I mean by nature only the aggregate action and product of many natural laws,—and by laws only the ascertained sequence of events.

Variation, vol. 1, 6

Difficulties

Long before having arrived at this part of my work, a crowd of difficulties will have occurred to the reader. Some of them are so grave that to this day I can never reflect on them without being staggered; but, to the best of my judgment, the greater number are only apparent, and those that are real are not, I think, fatal to my theory.

Origin 1859, 171

The eye to this day gives me a cold shudder, but when I think of the fine known gradations, my reason tells me I ought to conquer the cold shudder.

Darwin to Asa Gray,
[8 or 9 February 1860], DCP 2710

The sight of a feather in a peacock's tail, whenever I gaze at it, makes me sick!

Darwin to Asa Gray,
3 April [1860], DCP 2743

The foregoing remarks lead me to say a few words on the protest lately made by some naturalists, against the utilitarian doctrine that every detail of structure has been produced for the good of its possessor. They believe that very many structures have been created for beauty in the eyes of man, or for mere variety. This doctrine, if true, would be absolutely fatal to my theory.

Origin 1859, 199

Variability is governed by many unknown laws, more especially by that of correlation of growth. Something may be attributed to the direct action of the conditions of life. Something must be attributed to use and disuse. The final result is thus rendered infinitely complex.

Origin 1859, 43

As ants work by inherited instincts and by inherited tools or weapons, and not by acquired knowledge and manufactured instruments, a perfect division of labour could be effected with them only by the workers being sterile; for had they been fertile, they would have intercrossed, and their instincts and structure would have become blended. And nature has, as I believe, effected this admirable division of labour in the communities of ants, by the means of natural selection. But I am bound to confess, that, with all my faith in this principle, I should never have anticipated that natural selection could have been efficient in so high a degree, had not the case of these neuter insects convinced me of the fact.

Origin 1859, 242

Why then is not every geological formation and every stratum full of such intermediate links? Geology assuredly does not reveal any such finely graduated organic chain; and this, perhaps, is the most obvious and gravest objection which can be urged against my theory. The explanation lies, as I believe, in the extreme imperfection of the geological record.

Origin 1859, 280

If we admit that the geological record is imperfect in an extreme degree, then such facts as the record gives, support the theory of descent with modification. New species have come on the stage slowly and at successive intervals; and the amount of change, after equal intervals of time, is widely different in different groups. The extinction of species and of whole groups of species, which has played so conspicuous a part in the history of the organic world, almost inevitably follows on the principle of natural selection; for old forms will be supplanted by new and improved forms. Neither single species nor groups of species reappear when the chain of ordinary generation has once been broken.

Origin 1859, 475

In works on natural history rudimentary organs are generally said to have been created "for the sake of symmetry," or in order "to complete the scheme of nature;" but this seems to me no explanation, merely a restatement of the fact. Would it be thought sufficient to say that because planets revolve in elliptic courses round the sun, satellites follow the same course round the planets, for the sake of symmetry, and to complete the scheme of nature?

Origin 1859, 453

Looking to geographical distribution, if we admit that there has been during the long course of ages much migration from one part of the world to another, owing to former climatal and geographical changes and to the many occasional and unknown means of dispersal, then we can understand, on the theory of descent with modification, most of the great leading facts in Distribution. We can see why there should be so striking a parallelism in the distribution of organic beings throughout space, and in their geological succession throughout time; for in both cases the beings have been connected by the bond of ordinary generation, and the means of modification have been the same. We see the full meaning of the wonderful fact, which must have struck every traveller, namely, that on the same continent, under the most diverse conditions, under heat and cold, on mountain and lowland, on deserts and marshes, most of the inhabitants within each great class are plainly related; for they will generally be descendants of the same progenitors and early colonists.

Origin 1859, 476–77

The framework of bones being the same in the hand of a man, wing of a bat, fin of the porpoise, and leg of the horse,—the same number of vertebræ forming the neck of the giraffe and of the elephant,—and innumerable other such facts, at once explain themselves on the theory of descent with slow and slight successive modifications.

Origin 1859, 479

Nothing at first can appear more difficult to believe than that the more complex organs and instincts should have been perfected, not by means superior to, though analogous with, human reason, but by the accumulation of innumerable slight variations, each good for the individual possessor. Nevertheless, this difficulty, though appearing to our imagination insuperably great, cannot be considered real if we admit the following propositions, namely, —that gradations in the perfection of any organ or instinct, which we may consider, either do now exist or could have existed, each good of its kind,—that all organs and instincts are, in ever so slight a degree, variable,—and, lastly, that there is a struggle for existence leading to the preservation of each profitable deviation of structure or instinct. The truth of these propositions cannot, I think, be disputed.

Origin 1859, 459

Design and Free Will

The old argument of design in nature, as given by [William] Paley, which formerly seemed to me so conclusive, fails, now that the law of natural selection has been discovered. We can no longer argue that, for instance, the beautiful hinge of a bivalve shell must have been made by an intelligent being, like the hinge of a door by man. There seems to be no more design in the variability of organic beings and in the action of natural selection, than in the course which the wind blows.

Autobiography, 87

With respect to the theological view of the question; this is always painful to me.—I am bewildered.—I had no intention to write atheistically. But I own that I cannot see, as plainly as others do, & as I sh^d wish to do, evidence of design & beneficence on all sides of us. There seems to me too much misery in the world. I cannot persuade myself that a beneficent & omnipotent God would have designedly created the Ichneumonidæ with the express intention of their feeding within the living bodies of caterpillars, or that a cat should play with mice.

> Darwin to Asa Gray,
> 22 May [1860], DCP 2814

On the other hand I cannot anyhow be contented to view this wonderful universe & especially the nature of man, & to conclude that everything is the result of brute force. I am inclined to look at everything as resulting from designed laws, with the details, whether good or bad, left to the working out of what we may call chance. Not that this notion *at all* satisfies me. I feel most deeply that the whole subject is too profound for the human intellect. A dog might as well speculate on the mind of Newton.

> Darwin to Asa Gray,
> 22 May [1860], DCP 2814

No astronomer in showing how movements of Planets are due to gravity, thinks it necessary to say that the law of gravity was designed that the planets sh^d. pursue the courses which they pursue.—I cannot believe that there is a bit more interference by the Creator in the construction of each species, than in the course of the planets.

Darwin to Charles Lyell,
17 June [1860], DCP 2833

One word more on "designed laws'" & "undesigned results". I see a bird which I want for food, take my gun & kill it, I do this *designedly*.—An innocent & good man stands under tree & is killed by flash of lightning. Do you believe (& I really sh^d like to hear) that God *designedly* killed this man? Many or most persons do believe this; I can't & don't.

Darwin to Asa Gray,
3 July [1860], DCP 2855

Some one has sent us "Macmillan"; and I must tell you how much I admire your article; though at the same time I must confess that I could not clearly follow you in some parts, which probably is in main part due to my not being at all accustomed to metaphysical trains of thought. . . . The mind refuses to look at this universe, being what it is, without having been designed; yet, where one would most expect design, namely, in the structure of a sentient being, the more I think on the subject, the less I can see proof of design.

> Darwin to his niece Julia Wedgwood,
> 11 July [1861], *Life and Letters*,
> vol. 1, 313–14

Astronomers do not state that God directs the course of each comet & planet.—The view that each variation has been providentially arranged seems to me to make natural selection entirely superfluous, & indeed takes whole case of appearance of new species out of the range of science.

> Darwin to Charles Lyell,
> [1 August 1861], DCP 3230

If an architect were to rear a noble and commodious edifice, without the use of cut stone, by selecting from the fragments at the base of a precipice wedge-formed stones for his arches, elongated stones for his lintels, and flat stones for his roof, we should admire his skill and regard him as the paramount power. Now, the fragments of stone, though indispensable to the architect, bear to the edifice built by him the same relation which the fluctuating variations of each organic being bear to the varied and admirable structures ultimately acquired by its modified descendants. Can it be reasonably maintained that the Creator intentionally ordered, if we use the words in any ordinary sense, that certain fragments of rock should assume certain shapes so that the builder might erect his edifice?

Variation, vol. 2, 430–31

I have often felt much difficulty about the proper application of the terms, will, consciousness, and intention. Actions, which were at first voluntary, soon became habitual, and at last hereditary, and may then be performed even in opposition to the will. Although they often reveal the state of the mind, this result was not at first either intended or expected.

Expression, 357

Variation and Heredity

No one supposes that all the individuals of the same species are cast in the very same mould. These individual differences are highly important for us, as they afford materials for natural selection to accumulate, in the same manner as man can accumulate in any given direction individual differences in his domesticated productions.

Origin 1859, 45

I am strongly inclined to suspect that the most frequent cause of variability may be attributed to the male and female reproductive elements having been affected prior to the act of conception. Several reasons make me believe in this; but the chief one is the remarkable effect which confinement or cultivation has on the functions of the reproductive system.

Origin 1859, 8

The number and diversity of inheritable deviations of structure, both those of slight and those of considerable physiological importance, is endless. . . . No breeder doubts how strong is the tendency to inheritance: like produces like is his fundamental belief.

Origin 1859, 12

I have hitherto sometimes spoken as if the variations—so common and multiform in organic beings under domestication, and in a lesser degree in those in a state of nature—had been due to chance. This, of course, is a wholly incorrect expression, but it serves to acknowledge plainly our ignorance of the cause of each particular variation.

Origin 1859, 131

I venture to advance the hypothesis of Pangenesis, which implies that the whole organisation, in the sense of every separate atom or unit, reproduces itself. Hence ovules and pollen-grains,—the fertilised seed or egg, as well as buds,—include and consist of a multitude of germs thrown off from each separate atom of the organism.

Variation, vol. 2, 357–58

These granules for the sake of distinctness may be called cell-gemmules, or, as the cellular theory is not fully established, simply gemmules. They are supposed to be transmitted from the parents to the offspring, and are generally developed in the generation which immediately succeeds, but are often transmitted in a dormant state during many generations and are then developed.

Variation, vol. 2, 374

The existence of free gemmules is a gratuitous assumption, yet can hardly be considered as very improbable, seeing that cells have the power of multiplication through the self-division of their contents. . . . The gemmules in each organism must be thoroughly diffused; nor does this seem improbable considering their minuteness, and the steady circulation of fluids throughout the body.

Variation, vol. 2, 378, 379

This principle of Reversion is the most wonderful of all the attributes of Inheritance. . . . What can be more wonderful than that characters, which have disappeared during scores, or hundreds, or even thousands of generations, should suddenly reappear perfectly developed, as in the case of pigeons and fowls when purely bred, and especially when crossed; or as with the zebrine stripes on dun-coloured horses, and other such cases?

Variation, vol. 2, 372, 373

When we hear it said that a man carries in his constitution the seeds of an inherited disease, there is much literal truth in the expression.

Variation, vol. 2, 404

I wish I had known of these views of Hippocrates, before I had published, for they seem almost identical with mine—merely a change of terms—& an application of them to classes of facts necessarily unknown to this old philosopher. The whole case is a good illustration of how rarely anything is new.—The notion of pangenesis has been a wonderful relief to my mind, (as it has to some *few* others) for during long years I could not conceive any possible explanation of inheritance, development &c &c, or understand in the least in what reproduction by seeds & buds consisted. Hippocrates has taken the wind out of my sails, but I care very little about being forestalled

Darwin to William Ogle,
6 March [1868], DCP 5987

Pangenesis has very few friends, so let me beg you not to give it up lightly. It may be foolish parental affection, but it has thrown a flood of light on my mind in regard a great series of complex phenomena.

Darwin to T. H. Farrer,
29 October [1868], DCP 6435

In the earlier editions of my "Origin of Species" I probably attributed too much to the action of natural selection or the survival of the fittest. I have altered the fifth edition of the *Origin* so as to confine my remarks to adaptive changes of structure. I had not formerly sufficiently considered the existence of many structures which appear to be, as far as we can judge, neither beneficial nor injurious; and this I believe to be one of the greatest oversights as yet detected in my work.

Descent 1871, vol. 1, 152

I have lately i.e. in new Edit, of Origin been moderating my zeal, & attributing much more to mere useless variability.

Darwin to A. R. Wallace,
27 March [1869], DCP 6684

I am aware that my view [on pangenesis] is merely a provisional hypothesis or speculation; but until a better one be advanced, it may be serviceable by bringing together a multitude of facts which are at present left disconnected by any efficient cause. As [William] Whewell, the historian of the inductive sciences, remarks:—"Hypotheses may often be of service to science, when they involve a certain portion of incompleteness, and even of error."

Variation, vol. 2, 357

When, therefore, Mr. Galton concludes from the fact that rabbits of one variety, with a large proportion of the blood of another variety in their veins, do not produce mongrelised offspring, that the hypothesis of Pangenesis is false, it seems to me that his conclusion is a little hasty. His words are, "I have now made experiments of transfusion and cross circulation on a large scale in rabbits, and have arrived at definite results, negativing, in my opinion, beyond all doubt the truth of the doctrine of Pangenesis." If Mr. Galton could have proved that the reproductive elements were contained in the blood of the higher animals, and were merely separated or collected by the reproductive glands, he would have made a most important physiological discovery.

Darwin 1871, 503

Origin of Life

I believe that animals have descended from at most only four or five progenitors, and plants from an equal or lesser number. Analogy would lead me one step further, namely, to the belief that all animals and plants have descended from some one prototype. . . . probably all the organic beings which have ever lived on this earth have descended from some one primordial form, into which life was first breathed.

Origin 1859, 483–44

A celebrated author and divine [Charles Kingsley] has written to me that "he has gradually learnt to see that it is just as noble a conception of the Deity to believe that He created a few original forms capable of self-development into other and needful forms, as to believe that He required a fresh act of creation to supply the voids caused by the action of His laws."

Origin 1861, 525

ORIGIN OF LIFE

There is grandeur in this view of life, with its several powers, having been originally breathed by the Creator into a few forms or into one; and that, whilst this planet has gone cycling on according to the fixed law of gravity, from so simple a beginning endless forms most beautiful and most wonderful have been, and are being, evolved.

Origin 1861, 525

It will be some time before we see "slime, snot or protoplasm" (what an elegant writer) generating a new animal. But I have long regretted that I truckled to public opinion & used Pentateuchal term of creation, by which I really meant "appeared'" by some wholly unknown process.—It is mere rubbish thinking, at present, of origin of life; one might as well think of origin of matter.

Darwin to J. D. Hooker,
[29 Mar 1863], DCP 4065

But if (& oh what a big if) we could conceive in some warm little pond with all sorts of ammonia & phosphoric salts,—light, heat, electricity &c present, that a protein compound was chemically formed, ready to undergo still more complex changes, at the present day such matter wd be instantly devoured, or absorbed, which would not have been the case before living creatures were formed.

Darwin to J. D. Hooker,
1 February [1871], DCP 7471

Survival of the Fittest

I have been so repeatedly struck by the utter inability of numbers of intelligent persons to see clearly or at all, the self acting & necessary effects of Nat Selection, that I am led to conclude that the term itself & your mode of illustrating it, however clear & beautiful to many of us are yet not the best adapted to impress it on the general naturalist public. . . . I wish therefore to suggest to you the possibility of entirely avoiding this source of misconception in your great work, (if not now too late) & also in any future editions of the "Origin", and I think it may be done without difficulty & very effectually by adopting [Herbert] Spencer's term (which he generally uses in preference to Nat. Selection) viz. "Survival of the fittest."

> A. R. Wallace to Darwin,
> 2 July 1866, DCP 5140

I fully agree with all that you say on the advantages of H. Spencer's excellent expression of "the survival of the fittest." This however had not occurred to me till reading your letter.

It is, however, a great objection to this term that it cannot be used as a substantive governing a verb; & that this is a real objection I infer from H. Spencer continually using the words natural selection. I formerly thought, probably in an exaggerated degree, that it was a great advantage to bring into connection natural & artificial selection; this indeed led me to use a term in common, and I still think it some advantage.

Darwin to A. R. Wallace,
5 July [1866], DCP 5145

The term Natural Selection has now been so largely used abroad & at home that I doubt whether it could be given up, & with all its faults I should be sorry to see the attempt made. Whether it will be rejected must now depend "on the survival of the fittest."

Darwin to A.R. Wallace,
5 July [1866], DCP 5145

This preservation, during the battle for life, of varieties which possess any advantage in structure, constitution, or instinct, I have called Natural Selection; and Mr. Herbert Spencer has well expressed the same idea by the Survival of the Fittest. The term "natural selection" is in some respects a bad one, as it seems to imply conscious choice; but this will be disregarded after a little familiarity.

Variation, vol. 1, 6

The power of Selection, whether exercised by man, or brought into play under nature through the struggle for existence and the consequent survival of the fittest, absolutely depends on the variability of organic beings. Without variability nothing can be effected; slight individual differences, however, suffice for the work, and are probably the sole differences which are effective in the production of new species.

Variation, vol. 2, 192

Responses to *On the Origin of Species*

I am *infinitely* pleased & proud at the appearance of my child. . . . I am so glad that you were so good as to undertake the publication of my Book.

> Darwin to John Murray,
> [3 November 1859], DCP 2514

For myself I really think it is the most interesting book I ever read, & can only compare it to the first knowledge of chemistry, getting into a new world or rather behind the scenes. To me the geographical distribution I mean the relation of islands to continents is the most convincing of the proofs, & the relation of the oldest forms to the existing species. I dare say I dont feel enough the absence of varieties, but then I dont in the least know if every thing now living were fossilized whether the palæontologists could distinguish them. In fact the a priori reasoning is so entirely satisfactory to me that if the facts wont fit in, why so much the worse for the facts is my feeling.

Erasmus Alvey Darwin to Darwin,
23 November [1859], DCP 2545

I have read your book with more pain than pleasure. Parts of it I admired greatly; parts I laughed at till my sides were almost sore; other parts I read with absolute sorrow; because I think them utterly false & grievously mischievous—You have deserted—after a start in that tram-road of all solid physical truth—the true method of induction—& started up a machinery as wild I think as Bishop Wilkin's locomotive that was to sail with us to the Moon. Many of your wide conclusions are based upon assumptions which can neither be proved nor disproved. Why then express them in the language & arrangements of philosophical induction? . . . There is a moral or metaphysical part of nature as well as a physical. A man who denies this is deep in the mire of folly.

Adam Sedgwick to Darwin,
24 November 1859, DCP 2548

I have heard by round about channel that [John] Herschel says my Book "is the law of higgledy-pigglety".—What this exactly means I do not know, but it is evidently very contemptuous.—If true this is great blow & discouragement.

> Darwin to Charles Lyell,
> [10 December 1859], DCP 2575

Yesterday evening when I read the Times of previous day I was amazed to find a splendid Essay & Review of me. Who can the author be? I am intensely curious. It included a eulogium of me, which quite touched me, though I am not vain enough to think it all deserved. —The Author is a literary man & German scholar.—He has read my Book very attentively; but what is very remarkable, it seems that he is a profound naturalist. He knows my Barnacle book, & appreciates it too highly. —Lastly he writes & thinks with quite uncommon force & clearness; & what is even still rarer his writing is seasoned with most pleasant wit. . . . Certainly I should have said that there was only one man in England who could have written this Essay & that *you* were the man. But I suppose I am wrong, & that there is some hidden genius of great calibre.

> Darwin to T. H. Huxley,
> 28 December [1859], DCP 2611

Everybody has read Mr. Darwin's book, or, at least, has given an opinion upon its merits or demerits; pietists, whether lay or ecclesiastic, decry it with the mild railing which sounds so charitable; bigots denounce it with ignorant invective; old ladies of both sexes consider it a decidedly dangerous book, and even savants, who have no better mud to throw, quote antiquated writers to show that its author is no better than an ape himself; while every philosophical thinker hails it as a veritable Whitworth gun in the armoury of liberalism; and all competent naturalists and physiologists, whatever their opinions as to the ultimate fate of the doctrines put forth, acknowledge that the work in which they are embodied is a solid contribution to knowledge and inaugurates a new epoch in natural history.

T.H. Huxley, *Westminster Review*, 1860, 541

Extinguished theologians lie about the cradle of every science as the strangled snakes beside that of Hercules; and history records that whenever science and orthodoxy have been fairly opposed, the latter has been forced to retire from the lists, bleeding and crushed if not annihilated; scotched, if not slain.

T. H. Huxley, *Westminster Review*, 1860, 556

My reflection, when I first made myself master of the central idea of the *Origin*, was, "How extremely stupid not to have thought of that!"

T. H. Huxley, quoted in
Life and Letters, vol. 2, 197

Except some skill in the exposition of his opinions, and a moderate acquaintance with the results of recent inquiry, the author of the *Vestiges* added nothing to the "development theory" of Lamarck that could weigh with a mind trained to scientific investigation. . . . When we say that the conclusions announced by Mr. Darwin are such as, if established, would cause a complete revolution in the fundamental doctrines of natural history—and further, that although his theory is essentially distinct from the development theory of the *Vestiges of Creation*, it tends so far in the same direction as to trench upon the territory of established religious belief—we imply that his work is one of the most important that for a long time past have been given to the public. We have not been amongst the foremost to pass our judgment upon it, for it is a book—we say it deliberately—that will not bear to be dealt with lightly.

Review, Darwin's Origin of Species,
Saturday Review, 24 December 1859, 775

These are the most important original observations, recorded in the volume of 1859: they are, in our estimation, its real gems,—few indeed and far apart, and leaving the determination of the origin of species very nearly where the author found it. . . . Failing the adequacy of such observations, not merely to carry conviction, but to give a colour to the hypothesis, we were then left to confide in the superior grasp of mind, strength of intellect, clearness and precision of thought and expression, which raise one man so far above his contemporaries, as to enable him to discern in the common stock of facts, of coincidences, correlations and analogies in Natural History, deeper and truer conclusions than his fellow-labourers had been able to reach. These expectations, we must confess, received a check on perusing the first sentence in the book.

R. Owen 1860, 494, 495–96

I have just read the Edinburgh [Review], which without doubt is by Owen. It is extremely malignant, clever & I fear will be very damaging.

Darwin to Charles Lyell,
10 April [1860], DCP 2754

Since natural science deals only with secondary or natural causes, the scientific terms of a theory of derivation of species . . . must needs be the same to the theist as to the atheist. . . . Wherefore, Darwin's reticence about efficient cause does not disturb us. He considers only the scientific questions.

> Asa Gray 1860, 412

There has been a plethora of Reviews, & I am really quite sick of myself.

> Darwin to Charles Lyell,
> 10 April [1860], DCP 2754

I know not how, or to whom, to express fully my admiration of Darwin's book. To him it would seem flattery, to others self-praise; but I do honestly believe that with however much patience I had worked and experimented on the subject, I could never have approached the completeness of his book, its vast accumulation of evidence, its overwhelming argument, and its admirable tone and spirit. I really feel thankful that it has not been left to me to give the theory to the world.

> A.R. Wallace to H. W. Bates,
> quoted in Wallace 1905,
> vol. 1, 374

You must let me say how I admire the generous manner in which you speak of my Book: most persons would in your position have felt some envy or jealousy. How nobly free you seem to be of this common failing of mankind. — But you speak far too modestly of yourself;—you would, if you had had my leisure done the work just as well, perhaps better, than I have done it.

> Darwin to A. R. Wallace,
> 18 May 1860, DCP 2807

The Bishop of Oxford [Samuel Wilberforce] came out strongly against a theory which holds it possible that man may be descended from an ape,—in which protest he is sustained by Prof. Owen, Sir Benjamin Brodie, Dr. Daubeny, and the most eminent naturalists assembled at Oxford [for the 1860 meeting of the British Association for the Advancement of Science]. But others—conspicuous among these, Prof. Huxley—have expressed their willingness to accept, for themselves, as well as for their friends and enemies, all actual truths, even the last humiliating truth of a pedigree not registered in the Herald's College. The dispute has at least made Oxford uncommonly lively during the week.

> Report of the BAAS meeting,
> *Athenæum*, 7 July 1860, 19

Is it credible that all favourable varieties of turnips are tending to become men?

> Wilberforce 1860, 239

If I would rather have a miserable ape for a grandfather or a man highly endowed by nature and possessed of great means and influence, and yet who employs those faculties for the mere purpose of introducing ridicule into a grave scientific discussion—I unhesitatingly affirm my preference for the ape.

> T. H. Huxley to F. Dyster,
> 9 September 1860, quoted in
> Jensen 1988, 168

How durst you attack a live Bishop in that fashion? I am quite ashamed of you! Have you no reverence for fine lawn sleeves? By Jove, you seem to have done it well.

> Darwin to T. H. Huxley,
> [5 July 1860], DCP 2861

The battle rages furiously in U. States. [Asa] Gray says he was preparing a speech which would take 1 ½ hour to deliver, & which he "fondly hoped would be a stunner'". He is fighting splendidly & there seem to have been many discussions with [Louis] Agassiz & others at the meetings. Agassiz pities me much at being so deluded.

> Darwin to J. D. Hooker,
> 30 May [1860], DCP 2818

The criterion of a true theory consists in the facility with which it accounts for facts accumulated in the course of long-continued investigations and for which the existing theories afforded no explanation. It can certainly not be said that Darwin's theory will stand by that test.

> L. Agassiz 1860, 147

My book has stirred up the mud with a vengeance; & it will be a blessing to me if all my friends do not get to hate me. But I look at it as certain, if I had not stirred up the mud some one else would very soon; so that the sooner the battle is fought the sooner it will be settled,—not that the subject will be settled in our lives' times.

> Darwin to Asa Gray,
> 3 July [1860], DCP 2855

I shd. have been utterly smashed had it not been for you & three others.

> Darwin to T. H. Huxley,
> 3 July [1860], DCP 2854

Mr Darwin has given the world a new science, and his name should, in my opinion, stand above that of every philosopher of ancient or modern times. The force of admiration can no further go!

> A. R. Wallace to George Silk,
> 1 September 1860, quoted in
> Wallace 1905, vol. 1, 373

Dr. Whewell dissented in a practical manner for some years, by refusing to allow a copy of the "Origin of Species" to be placed in the Library of Trinity College [Cambridge].

> F. Darwin in *Life and Letters*, v. 2, 261n.

I reckon Darwin's book to be an utterly *unphilosophical* one.

> William Whewell, letter to J. D. Forbes,
> 24 July 1860, quoted in *Dictionary
> of Scientific Quotations*, 619

For Heaven sake don't write an anti-Darwinian article; you would do it so confoundedly well. . . . I shall always think those early Reviews, almost entirely yours, did the subject an *enormous* service.

> Darwin to T. H. Huxley,
> 22 November [1860], DCP 2994

Am I satyr or man?
  Pray tell me who can,
And settle my place in the scale.
  A man in ape's shape,
  An anthropoid ape,
Or monkey deprived of his tail?

> Anon, *Punch*, 18 May 1861

When I came to the conclusion that after all Lamarck was going to be shown to be right, that we must "go the whole orang," I re-read his book, and remembering when it was written, I felt I had done him injustice.

> Charles Lyell to Darwin,
> 15 March 1863, DCP 4041

The question is this: Is man an ape or an angel? Now I am on the side of the angels.

> Benjamin Disraeli, Speech, quoted in
> *Punch*, 10 December 1864

I shall never forget that meeting of the combined sections of the British Association when at Oxford 1860, where Admiral Fitzroy expressed his sorrows for having given you the opportunities of collecting facts for such a shocking theory as yours.

> J. V. Carus to Darwin,
> 15 November 1866, DCP 5282

I received, 2 or 3 days ago, a French translation of the "Origin," by a Mlle. Royer, who must be one of the cleverest & oddest women in Europe: is ardent Deist, & hates Christianity, & declares that natural selection & the struggle for life will explain all morality, nature of man, politics, &c. &c.!

>   Darwin to Asa Gray,
>   10–20 June [1862], DCP 3595

It is remarkable how Darwin rediscovers among beasts and plants the society of England, with its division of labour, competition, opening up of new markets, inventions, and the Malthusian struggle for existence.

>   Karl Marx to F. Engels, 18 June 1862,
>   Marx 1975–2004, vol. 41, 381

It is unreasonable to accuse Mr. Darwin (as has been done) of violating the rules of induction. The rules of induction are concerned with the condition of proof. Mr. Darwin has never pretended that his doctrine was proved. He was not bound by the rules of induction but by those of hypothesis. And these last have seldom been more completely fulfilled. He has opened a path of inquiry full of promise, the results of which none can foresee.

>   Mill 1862, vol. 2, 180

I have never, I think, in my life, been so deeply interested by any geological discussion. I now first begin to see what a million means, and I feel quite ashamed of myself at the silly way in which I have spoken of millions of years. I was formerly a great believer in the power of the sea in denudation and this was perhaps natural, as most of my geological work was done near sea coasts, and on islands. . . . How often I have speculated in vain on the origin of the vallies in the chalk platform round this place [Down House, Kent], but now all is clear. I thank you cordially for having cleared so much mist from before my eyes.

> Darwin to James Croll,
> 19 September 1868, DCP 6380

Fleming Jenkins [Fleeming Jenkin] has given me much trouble, but has been of more real use to me, than any other Essay or Review.

> Darwin to J. D. Hooker,
> 16 January [1869], DCP 6557

I am greatly troubled at the short duration of the world according to Sir W. Thompson, for I require for my theoretical views a very long period before the Cambrian [geological] formation.

> Darwin to James Croll,
> 31 January [1869], DCP 6585

Hardly any point gave me so much satisfaction when I was at work on the *Origin*, as the explanation of the wide difference in many classes between the embryo and the adult animal, and of the close resemblance of the embryos within the same class. No notice of this point was taken, as far as I remember, in the early reviews of the *Origin*, and I recollect expressing my surprise on this head in a letter to Asa Gray.

Autobiography, 125

The conclusion of the whole matter is, that the denial of design in nature is virtually the denial of God. . . . We have thus arrived at the answer to our question, What is Darwinism? It is Atheism.

Hodge 1874, 177

Some of my critics have said, "Oh, he is a good observer, but has no power of reasoning." I do not think that this can be true, for the *Origin of Species* is one long argument from the beginning to the end, and it has convinced not a few able men. No one could have written it without having some power of reasoning.

Autobiography, 140

Botany

I must beg *sometime* for a single sentence about the Galapagos plants. viz what percentage are (as far as is known) peculiar to the Archipelago? you have already told me that the plants have a S. American physionomy. And how far the collections bear out or contradict the notion of the different islands, having in some instances representative & different species.

Darwin to J. D. Hooker,
[16 April 1845], DCP 848

[I am] a man who hardly knows a daisy from a Dandelion.

Darwin to J. D. Hooker,
[3 September 1846], DCP 996

Miss Thorley & I are doing a *little Botanical work* (!) for our amusement, & it does amuse me very much, viz making a collection of all the plants, which grow in a field, which has been allowed to run waste for 15 years . . . & we are also collecting all the plants in an adjoining & *similar* but cultivated field; just for the fun of seeing what plants have arrived or dyed out. Hereafter we shall want a bit of help in naming puzzlers.—How dreadfully difficult it is to name plants.

> Darwin to J. D. Hooker,
> 5 June [1855], DCP 1693

I have just made out my first Grass, hurrah! hurrah! I must confess that Fortune favours the bold, for as good luck wd have it, it was the easy Anthoxanthum odoratum: nevertheless it is a great discovery; I never expected to make out a grass in all my life. So Hurrah. It has done my stomach surprising good.

> Darwin to J. D. Hooker,
> 5 June [1855], DCP1693

I have been very lucky & have now examined almost every British Orchid fresh. . . . I cannot fancy anything more perfect than the many curious contrivances.

> Darwin to J. D. Hooker,
> 19 June [1861], DCP 3190

The object of the following work is to show that the contrivances by which Orchids are fertilised, are as varied and almost as perfect as any of the most beautiful adaptations in the animal kingdom; and, secondly, to show that these contrivances have for their main object the fertilisation of each flower. . . . This treatise affords me also an opportunity of attempting to show that the study of organic beings may be as interesting to an observer who is fully convinced that the structure of each is due to secondary laws, as to one who views every trifling detail of structure as the result of the direct interposition of the Creator.

Orchids, 1

In my examination of Orchids, hardly any fact has so much struck me as the endless diversity of structure,—the prodigality of resources, —for gaining the very same end, namely, the fertilisation of one flower by the pollen of another. The fact to a certain extent is intelligible on the principle of natural selection. As all the parts of a flower are co-ordinated, if slight variations in any one part are preserved from being beneficial to the plant, then the other parts will generally have to be modified in some corresponding manner.

Orchids, 348–49

[James] Bateman has just sent me a lot of orchids with the Angræcum sesquipedale: do you know its marvellous nectary 11 ½ inches long, with nectar only at the extremity. What a proboscis the moth that sucks it, must have! It is a very pretty case.

> Darwin to J. D. Hooker,
> 30 January [1862], DCP 3421

No one else has perceived that my chief interest in my orchid book has been that it was a "flank movement" on the enemy.

> Darwin to Asa Gray,
> 23[–24] July [1862], DCP 3662

In the summer of 1860 I was idling and resting near Hartfield [Sussex], where two species of Drosera abound; and I noticed that numerous insects had been entrapped by the leaves. I carried home some plants, and on giving them insects saw the movements of the tentacles, and this made me think it probable that the insects were caught for some special purpose. . . . The fact that a plant should secrete, when properly excited, a fluid containing an acid and ferment, closely analogous to the digestive fluid of an animal, was certainly a remarkable discovery.

> *Autobiography*, 132–33

At this present moment I care more about Drosera than the origin of all the species in the world.

> Darwin to Charles Lyell,
> 14 November [1860], DCP 2565

By Jove I sometimes think Drosera is a disguised animal!

> Darwin to J. D. Hooker,
> 4 December [1860], DCP 3008

I write now, because the new Hothouse is ready & I long to stock it, just like a schoolboy.—Could you tell me pretty soon what plants you can give me; & then I shall know what to order. And do advise me how I had better get such plants as you can spare. Would it do to send my tax-cart early in morning, on a day that was not frosty, lining the cart with mats; & arriving here before night.

> Darwin to J. D. Hooker,
> 15 February [1863], DCP 3986

The only approach to work which I can do is to look at tendrils & climbers, this does not distress my weakened Brain.

> Darwin to J. D. Hooker,
> [27 January 1864], DCP 4398

I am glad to hear the Abutilon is a new species, & I am honoured by its name [*Abutilon darwinii*]. I do not know its habitat, but strongly suspect that it must be St. Catharina [Brazil]. The plant flourished & flowered profusely in my cool hot-house.—It seems to like heat. It offers an instance, of which I have known others, of being during the early part of the flowering season quite sterile with pollen from the same plant, though fertile with the pollen of any other plant, though later in the season it becomes capable of self-fertilisation.

Darwin to J. D. Hooker,
23 July [1871], DCP 7878

I do not think anything in my scientific life has given me so much satisfaction as making out the meaning of the structure of these plants [Primula]. . . . After some additional experiment, it became evident that the two forms, though both were perfect hermaphrodites, bore almost the same relation to one another as do the two sexes of an ordinary animal.

Autobiography, 126–27

PART 4

Mankind

Caricature, *The Hornet*, 22 March 1871. Reproduced with permission from Special Collections, University College London.

Human Origins

As soon as I had become, in the year 1837 or 1838, convinced that species were mutable productions, I could not avoid the belief that man must come under the same law. Accordingly I collected notes on the subject for my own satisfaction, and not for a long time with any intention of publishing. Although in the *Origin of Species*, the derivation of any particular species is never discussed, yet I thought it best, in order that no honourable man should accuse me of concealing my views, to add that by the work in question "light would be thrown on the origin of man and his history."
Autobiography, 130

You ask whether I shall discuss "man";—I think I shall avoid whole subject, as so surrounded with prejudices, though I fully admit that it is the highest & most interesting problem for the naturalist.

Darwin to A. R. Wallace,
22 December 1857, DCP 2192

I am sorry to say that I have no "consolatory view" on the dignity of man; I am content that man will probably advance & care not much whether we are looked at as mere savages in a remotely distant future.

> Darwin to Charles Lyell,
> 4 May [1860], DCP 2782

I was partly led to do this by having been taunted that I concealed my views, but chiefly from the interest which I had long taken in the subject.

> Darwin to Alphonse de Candolle,
> 6 July 1868, DCP 6269

The mental requirements of the lowest savages, such as the Australians or the Andaman islanders, are very little above those of many animals. . . . How then was an organ [the brain] developed far beyond the needs of its possessor? Natural Selection could only have endowed the savage with a brain a little superior to that of an ape, whereas he actually possesses one but very little inferior to that of the average members of our learned societies.

> Wallace 1869, 391–92

I hope you have not murdered too completely your own & my child.
> Darwin to A. R. Wallace,
> 27 March [1869], DCP 6684

I differ grievously from you, & I am very sorry for it. I can see no necessity for calling in an additional & proximate cause in regard to Man. But the subject is too long for a letter. I have been particularly glad to read y^r discussion because I am now writing & thinking much about man.
> Darwin to A. R. Wallace,
> 14 April 1869, DCP 6706

During many years I collected notes on the origin or descent of man, without any intention of publishing on the subject, but rather with the determination not to publish, as I thought that I should thus only add to the prejudices against my views. . . . Now the case wears a wholly different aspect. . . . The greater number [of naturalists] accept the agency of natural selection; though some urge, whether with justice the future must decide, that I have greatly overrated its importance. Of the older and honoured chiefs in natural science, many unfortunately are still opposed to evolution in every form.
> *Descent* 1871, vol. 1, 2

It might be intelligible that a man's tail should waste away when he had no longer occasion to wag it, though I should have thought that savages would still have found it useful in tropical climates to brush away insects. . . . The arguments in the sheets [of *Descent of Man*] you have sent me appear to me to be little better than drivel.

> Whitwell Elwin to John Murray,
> 21 September 1870, John Murray
> Archives, National Library of Scotland

Man bears in his bodily structure clear traces of his descent from some lower form.
> *Descent* 1871, vol. 1, 34

The early progenitors of man were no doubt once covered with hair, both sexes having beards; their ears were pointed and capable of movement; and their bodies were provided with a tail, having the proper muscles. Their limbs and bodies were also acted on by many muscles which now only occasionally reappear, but are normally present in the Quadrumana. . . . The foot, judging from the condition of the great toe in the fœtus, was then prehensile; and our progenitors, no doubt, were arboreal in their habits, frequenting some warm, forest-clad land. The males were provided with great canine teeth, which served them as formidable weapons.

Descent 1871, vol. 1, 206–7

In a series of forms graduating insensibly from some ape-like creature to man as he now exists, it would be impossible to fix on any definite point when the term "man" ought to be used.

Descent 1871, vol. 1, 235

As monkeys certainly understand much that is said to them by man, and as in a state of nature they utter signal-cries of danger to their fellows, it does not appear altogether incredible, that some unusually wise ape-like animal should have thought of imitating the growl of a beast of prey, so as to indicate to his fellow monkeys the nature of the expected danger. And this would have been a first step in the formation of a language.

Descent 1871, vol. 1, 57

The main conclusion arrived at in this work, namely that man is descended from some lowly-organised form, will, I regret to think, be highly distasteful to many persons. But there can hardly be a doubt that we are descended from barbarians. The astonishment which I felt on first seeing a party of Fuegians on a wild and broken shore will never be forgotten by me, for the reflection at once rushed into my mind—such were our ancestors. These men were absolutely naked and bedaubed with paint, their long hair was tangled, their mouths frothed with excitement, and their expression was wild, startled, and distrustful. They possessed hardly any arts, and like wild animals lived on what they could catch; they had no government, and were merciless to every one not of their own small tribe. He who has seen a savage in his native land will not feel much shame, if forced to acknowledge that the blood of some more humble creature flows in his veins.

Descent 1871, vol. 2, 404

I have given the evidence to the best of my ability; and we must acknowledge, as it seems to me, that man with all his noble qualities, with sympathy which feels for the most debased, with benevolence which extends not only to other men but to the humblest living creature, with his god-like intellect which has penetrated into the movements and constitution of the solar system—with all these exalted powers—Man still bears in his bodily frame the indelible stamp of his lowly origin.

Descent 1871, vol. 2, 405

Mr. Darwin's conclusions may be correct, but we feel we have now indeed a right to demand that they shall be proved before we assent to them; and that since what Mr. Darwin before declared "*must* be," he now admits not only to be unnecessary but untrue, we may justly regard with extreme distrust the numerous statements and calculations which, in the "Descent of Man," are avowedly recommended by a mere "*may* be."

George St. J. Mivart, 1871, 52

Altogether the book [*Descent of Man*], I think, as yet has been very successful, & I have been hardly at all abused. Several reviewers speak of the lucid vigorous style etc.—Now I know how much I owe to you in this respect, which includes arrangement, not to mention still more important aids in the reasoning. Therefore I wish to give you some little memorial costing about 25 or 50£, to keep in memory of the book, over which you took such immense trouble. I have consulted Mamma, but we cannot think what you would like, & she with her accustomed wisdom advised me to lay the case before you & let you decide how you like. . . . By the way, I have had hardly any letters about "the Descent" worth keeping for you, except one from a Welshman abusing me as an old ape with a hairy face & thick skull.

We shall be heartily glad to see you home again. Goodbye my very dear coadjutor & fellow-labourer, Your affec^(ate). father. Ch. Darwin.

> Darwin to Henrietta Darwin,
> 20 March 1871, DCP 7605

Race

When two races of men meet they act precisely like two species of animals,—they fight, eat each other, bring diseases to each other &c., but then comes the more deadly struggle, namely which have the best fitted organizations, or instincts (ie intellect in man) to gain the day. . . . Man acts on & is acted on by the organic and inorganic agents of this earth like every other animal.

Notebook E, 63, 65

I suspect that a sort of sexual selection has been the most powerful means of changing the races of man. I can shew that the diff^t races have a widely diff^t standard of beauty. Among savages the most powerful men will have the pick of the women & they will generally leave the most descendants.

Darwin to A. R. Wallace,
28 [May 1864], DCP 4510

Probably you are right on all the points you touch on except as I think about sexual selection which I will not give up.... It is an awful stretcher to believe that a Peacock's tail was thus formed, but believing it, I believe in the same principle somewhat modified applied to man.

> Darwin to A. R. Wallace,
> 15 June [1864], DCP 4535

Man tends to multiply at so rapid a rate that his offspring are necessarily exposed to a struggle for existence, and consequently to natural selection. He has given rise to many races, some of which are so different that they have often been ranked by naturalists as distinct species.

> *Descent* 1871, vol. 1, 185

Although the existing races of man differ in many respects, as in colour, hair, shape of skull, proportions of the body, &c., yet if their whole organisation be taken into consideration they are found to resemble each other closely in a multitude of points.

> *Descent* 1871, vol. 1, 231–32

The belief that there exists in man some close relation between the size of the brain and the development of the intellectual faculties is supported by the comparison of the skulls of savage and civilised races, of ancient and modern people.

Descent 1871, vol. 1, 145

I do not intend to assert that sexual selection will account for all the differences between the races.

Descent 1871, vol. 1, 249

The strongest and most vigorous men,—those who could best defend and hunt for their families, and during later times the chiefs or head-men,—those who were provided with the best weapons and who possessed the most property, such as a larger number of dogs or other animals, would have succeeded in rearing a greater average number of offspring, than would the weaker, poorer and lower members of the same tribes. There can, also, be no doubt that such men would generally have been able to select the more attractive women.

Descent 1871, vol. 2, 368–69

It would be an inexplicable circumstance, if the selection of the more attractive women by the more powerful men of each tribe, who would rear on an average a greater number of children, did not after the lapse of many generations modify to a certain extent the character of the tribe.

Descent 1871, vol. 2, 369

Sexual Selection

In the same manner as man can improve the breed of his game-cocks by the selection of those birds which are victorious in the cockpit, so it appears that the strongest and most vigorous males, or those provided with the best weapons, have prevailed under nature, and have led to the improvement of the natural breed or species.

Descent 1871, vol. 1, 258

When the sexes differ in colour or in other ornaments, the males with rare exceptions are the most highly decorated, either permanently or temporarily during the breeding-season. They sedulously display their various ornaments, exert their voices, and perform strange antics in the presence of the females. Even well-armed males, who, it might have been thought, would have altogether depended for success on the law of battle, are in most cases highly ornamented; and their ornaments have been acquired at the expense of some loss of power. In other cases ornaments have been acquired, at the cost of increased risk from birds and beasts of prey.

Descent 1871, vol. 2, 123

All animals present individual differences, and as man can modify his domesticated birds by selecting the individuals which appear to him the most beautiful, so the habitual or even occasional preference by the female of the more attractive males would almost certainly lead to their modification; and such modifications might in the course of time be augmented to almost any extent, compatible with the existence of the species.

Descent 1871, vol. 2, 124

The peacock with his long train appears more like a dandy than a warrior, but he sometimes engages in fierce contests.

Descent 1871, vol. 2, 46

Many will declare that it is utterly incredible that a female bird should be able to appreciate fine shading and exquisite patterns. It is undoubtedly a marvellous fact that she should possess this almost human degree of taste, though perhaps she admires the general effect rather than each separate detail. He who thinks that he can safely gauge the discrimination and taste of the lower animals, may deny that the female Argus pheasant can appreciate such refined beauty; but he will then be compelled to admit that the extraordinary attitudes assumed by the male during the act of courtship, by which the wonderful beauty of his plumage is fully displayed, are purposeless; and this is a conclusion which I for one will never admit.

Descent 1871, vol. 2, 93

As negroes, as well as savages in many parts of the world, paint their faces with red, blue, white, or black bars,—so the male mandrill of Africa appears to have acquired his deeply-furrowed and gaudily-coloured face from having been thus rendered attractive to the female. No doubt it is to us a most grotesque notion that the posterior end of the body should have been coloured for the sake of ornament even more brilliantly than the face; but this is really not more strange than that the tails of many birds should have been especially decorated.

Descent 1871, vol. 2, 296

Man is more powerful in body and mind than woman, and in the savage state he keeps her in a far more abject state of bondage than does the male of any other animal; therefore it is not surprising that he should have gained the power of selection.

Descent 1871, vol. 2, 371

As far as sexual selection is concerned, all that is required is that choice should be exerted before the parents unite, and it signifies little whether the unions last for life or only for a season.

Descent 1871, vol. 2, 360

If an inhabitant of another planet were to behold a number of young rustics at a fair, courting and quarrelling over a pretty girl, like birds at one of their places of assemblage, he would be able to infer that she had the power of choice only by observing the eagerness of the wooers to please her, and to display their finery.

Descent 1871, vol. 2, 122

My conviction of the power of sexual selection remains unshaken; but it is probable, or almost certain, that several of my conclusions will hereafter be found erroneous; this can hardly fail to be the case in the first treatment of a subject. When naturalists have become familiar with the idea of sexual selection, it will, as I believe, be much more largely accepted; and it has already been fully and favourably received by several capable judges.

Descent 1874, vol. 1, vi

Morality

It has interested me much to see how differently two men may look at the same points, though I fully feel how presumptuous it sounds to put myself even for a moment in the same bracket with Kant;—the one man a great philosopher looking exclusively into his own mind, the other a degraded wretch looking from the outside thro' apes & savages at the moral sense of mankind.

> Darwin to Frances Power Cobbe,
> 23 March [1870], DCP 7149

Of all the differences between man and the lower animals, the moral sense or conscience is by far the most important.

> *Descent* 1871, vol. 1, 70

If, for instance, to take an extreme case, men were reared under precisely the same conditions as hive-bees, there can hardly be a doubt that our unmarried females would, like the worker-bees, think it a sacred duty to kill their brothers, and mothers would strive to kill their fertile daughters; and no one would think of interfering.

Descent 1871, vol. 1, 73

A moral being is one who is capable of comparing his past and future actions or motives, and of approving or disapproving of them. We have no reason to suppose that any of the lower animals have this capacity; therefore when a monkey faces danger to rescue its comrade, or takes charge of an orphan-monkey, we do not call its conduct moral. But in the case of man, who alone can with certainty be ranked as a moral being, actions of a certain class are called moral, whether performed deliberately after a struggle with opposing motives, or from the effects of slowly-gained habit, or impulsively through instinct.

Descent 1871, vol. 1, 88–89

As soon, however, as marriage, whether polygamous or monogamous, becomes common, jealousy will lead to the inculcation of female virtue; and this being honoured will tend to spread to the unmarried females. How slowly it spreads to the male sex we see at the present day. Chastity eminently requires self-command; therefore it has been honoured from a very early period in the moral history of civilised man. As a consequence of this, the senseless practice of celibacy has been ranked from a remote period as a virtue.

Descent 1871, vol. 1, 96

I believe that any such practices [contraception] would in time spread to unsound women & would destroy chastity, on which the family bond depends; & the weakening of this bond would be the greatest of all possible evils to mankind.

Darwin to Charles Bradlaugh,
6 June 1877, quoted in
Peart and Levy 2008, 348

Intellect

There is no fundamental difference between man and the higher mammals in their mental faculties.

Descent 1871, vol. 1, 35

There can be no doubt that the difference between the mind of the lowest man and that of the highest animal is immense.

Descent 1871, vol. 1, 104

The mental powers in some early progenitor of man must have been more highly developed than in any existing ape, before even the most imperfect form of speech could have come into use; but we may confidently believe that the continued use and advancement of this power would have reacted on the mind by enabling and encouraging it to carry on long trains of thought.

Descent 1871, vol. 1, 57

At the present day civilised nations are everywhere supplanting barbarous nations, excepting where the climate opposes a deadly barrier; and they succeed mainly, though not exclusively, through their arts, which are the products of the intellect. It is, therefore, highly probable that with mankind the intellectual faculties have been gradually perfected through natural selection.

Descent 1871, vol. 1, 160

We may conclude that the greater size, strength, courage, pugnacity, and even energy of man, in comparison with the same qualities in woman, were acquired during primeval times, and have subsequently been augmented, chiefly through the contests of rival males for the possession of the females. The greater intellectual vigour and power of invention in man is probably due to natural selection combined with the inherited effects of habit, for the most able men will have succeeded best in defending and providing for themselves, their wives and offspring.

Descent 1871, vol. 2, 382–83

Woman seems to differ from man in mental disposition, chiefly in her greater tenderness and less selfishness; and this holds good even with savages. . . . Man is the rival of other men; he delights in competition, and this leads to ambition which passes too easily into selfishness. These latter qualities seem to be his natural and unfortunate birthright. It is generally admitted that with woman the powers of intuition, of rapid perception, and perhaps of imitation, are more strongly marked than in man; but some, at least, of these faculties are characteristic of the lower races, and therefore of a past and lower state of civilisation.

Descent 1871, vol. 2, 326–27

The chief distinction in the intellectual powers of the two sexes is shewn by man attaining to a higher eminence, in whatever he takes up, than woman can attain—whether requiring deep thought, reason, or imagination, or merely the use of the senses and hands. If two lists were made of the most eminent men and women in poetry, painting, sculpture, music—comprising composition and performance—history, science, and philosophy, with half-a-dozen names under each subject, the two lists would not bear comparison.

Descent 1871, vol. 2, 327

In order that woman should reach the same standard as man, she ought, when nearly adult, to be trained to energy and perseverance, and to have her reason and imagination exercised to the highest point; and then she would probably transmit these qualities chiefly to her adult daughters. The whole body of women, however, could not be thus raised, unless during many generations the women who excelled in the above robust virtues were married, and produced offspring in larger numbers than other women.

Descent 1871, vol. 2, 329

I do not think I ever in all my life read anything more interesting & original [F. Galton *Hereditary Genius*, 1869]. . . . You have made a convert of an opponent in one sense, for I have always maintained that, excepting fools, men did not differ much in intellect, only in zeal & hard work; & I still think there is an eminently important difference. I congratulate you on producing what I am convinced will prove a memorable work

Darwin to Francis Galton,
23 December [1869], DCP 7032

I hope that it will not bore you to read the few accompanying pages, and in the middle you will find a few sentences with a sort of definition of, or rather discussion on, intelligence. I am altogether dissatisfied with it. I tried to observe what passed in my own mind when

I did the work of a worm. If I come across a professed metaphysician, I will ask him to give me a more technical definition, with a few big words about the abstract, the concrete, the absolute, and the infinite; but seriously, I should be grateful for any suggestions, for it will hardly do to assume that every fool knows what "intelligent" means.

Darwin to G. J. Romanes,
7 March 1881, More Letters, vol. 2,
213–14

I am inclined to agree with Francis Galton in believing that education and environment produce only a small effect on the mind of any one, and that most of our qualities are innate.

Autobiography, 43

Instincts

If it can be shown that instincts do vary ever so little, then I can see no difficulty in natural selection preserving and continually accumulating variations of instinct to any extent that may be profitable. It is thus, as I believe, that all the most complex and wonderful instincts have originated.

Origin 1859, 209

By what steps the instinct of F. sanguinea [the slave-making ant] originated I will not pretend to conjecture. But as ants, which are not slave-makers, will, as I have seen, carry off pupæ of other species, if scattered near their nests, it is possible that pupæ originally stored as food might become developed; and the ants thus unintentionally reared would then follow their proper instincts, and do what work they could. If their presence proved useful to the species which had seized them—if it were more advantageous to this species to capture workers than to procreate them—the habit of collecting pupæ originally for food might by natural selection be strengthened and rendered permanent for the very different purpose of raising slaves.

Origin 1859, 223–24

Thus, as I believe, the most wonderful of all known instincts, that of the hive-bee [in constructing hexagonal cells in honeycombs], can be explained by natural selection having taken advantage of numerous, successive, slight modifications of simpler instincts; natural selection having by slow degrees, more and more perfectly, led the bees to sweep equal spheres at a given distance from each other in a double layer, and to build up and excavate the wax along the planes of intersection. The bees, of course, no more knowing that they swept their spheres at one particular distance from each other, than they know what are the several angles of the hexagonal prisms and of the basal rhombic plates. The motive power of the process of natural selection having been economy of wax; that individual swarm which wasted least honey in the secretion of wax, having succeeded best, and having transmitted by inheritance its newly acquired economical instinct to new swarms, which in their turn will have had the best chance of succeeding in the struggle for existence.

Origin 1859, 235

No one supposes that one of the lower animals reflects whence he comes or whither he goes, —what is death or what is life, and so forth. But can we feel sure that an old dog with an excellent memory and some power of imagination, as shewn by his dreams, never reflects on his past pleasures in the chase? and this would be a form of self-consciousness. On the other hand, as [Georg] Büchner has remarked, how little can the hard-worked wife of a degraded Australian savage, who uses hardly any abstract words and cannot count above four, exert her self-consciousness, or reflect on the nature of her own existence.

Descent 1871, vol. 1, 62

The very essence of an instinct is that it is followed independently of reason.

Descent 1871, vol. 1, 100

Some intelligent actions, after being performed during several generations, become converted into instincts and are inherited, as when birds on oceanic islands learn to avoid man. . . . The greater number of the more complex instincts appear to have been gained in a wholly different manner, through the natural selection of variations of simpler instinctive actions. Such variations appear to arise from the same unknown causes acting on the cerebral organisation, which induce slight variations or individual differences in other parts of the body; and these variations, owing to our ignorance, are often said to arise spontaneously.

Descent 1874, 67

Expression of the Emotions

My first child was born on December 27th, 1839, and I at once commenced to make notes on the first dawn of the various expressions which he exhibited, for I felt convinced, even at this early period, that the most complex and fine shades of expression must all have had a gradual and natural origin.

Autobiography, 131–32

Give Mrs Huxley the enclosed [a questionnaire on facial expressions] & ask her to look out (for no 5) when one of her children is struggling & just going to burst out crying. A dear young lady near here, plagued a very young child for my sake, till it cried, & saw the eyebrows for a second or two beautifully oblique, just before the torrent of tears began.

Darwin to T. H. Huxley,
30 January [1868], DCP 5817

When the Callithrix sciureus screams violently does it wrinkle up the skin round the eyes like a Baby always does? When thus screaming do the eyes become suffused with moisture? Will

you ask Sutton [a keeper at the Zoological Gardens, London] to observe carefully.
—Could you make it scream without hurting it much?

Darwin to Abraham D. Bartlett
5 January [1870], DCP 7072

Most of the more complex emotions are common to the higher animals and ourselves. Every one has seen how jealous a dog is of his master's affection, if lavished on any other creature; and I have observed the same fact with monkeys. This shows that animals not only love, but have the desire to be loved. Animals manifestly feel emulation. They love approbation or praise; and a dog carrying a basket for his master exhibits in a high degree self-complacency or pride. There can, I think, be no doubt that a dog feels shame, as distinct from fear, and something very like modesty when begging too often for food. A great dog scorns the snarling of a little dog, and this may be called magnanimity.

Descent 1871, vol. 1, 41–42

I had intended to give only a chapter on the subject in the *Descent of Man*, but as soon as I began to put my notes together, I saw that it would require a separate Treatise.

Autobiography, 131

A young female chimpanzee, in a violent passion, presented a curious resemblance to a child in the same state. She screamed loudly with widely open mouth, the lips being retracted so that the teeth were fully exposed. She threw her arms wildly about, sometimes clasping them over her head.

Expression, 140

As the lacrymal glands are remarkably free from the control of the will, they would be eminently liable still to act, thus betraying, though there were no other outward signs, the pathetic thoughts which were passing through the person's mind.

Expression, 175

Many years ago, in the Zoological Gardens, I placed a looking-glass on the floor before two young orangs, who, as far as it was known, had never before seen one. At first they gazed at their own images with the most steady surprise, and often changed their point of view. They then approached close and protruded their lips towards the image, as if to kiss it, in exactly the same manner as they had previously done towards each other, when first placed, a few days before, in the same room. They next made all sorts of grimaces, and put themselves in various attitudes before the mirror; they pressed and rubbed the surface; they placed their hands at different distances behind it; looked behind it; and finally seemed almost frightened, started a little, became cross, and refused to look any longer.

Expression, 142

With one of my own infants, from his eighth day and for some time afterwards, I often observed that the first sign of a screaming-fit, when it could be observed coming on gradually, was a little frown, owing to the contraction of the corrugators of the brows; the capillaries of the naked head and face becoming at the same time reddened with blood. As soon as the screaming-fit actually began, all the muscles round the eyes were strongly contracted, and the mouth widely opened in the manner above described; so that at this early period the features assumed the same form as at a more advanced age.

Expression, 151–52

I have endeavoured to show in considerable detail that all the chief expressions exhibited by man are the same throughout the world. This fact is interesting, as it affords a new argument in favour of the several races being descended from a single parent-stock, which must have been almost completely human in structure, and to a large extent in mind, before the period at which the races diverged from each other. . . . it seems to me improbable in the highest degree that so much similarity, or rather identity of structure, could have been acquired by independent means.

Expression, 361

Human Society

It is necessary to leave England, & see distant Colonies of various nations, to know what wonderful people the English are.

Darwin to E. C. Darwin,
14 February 1836, DCP 298

The perfect equality among the individuals composing these [Fuegian] tribes, must for a long time retard their civilization. As we see those animals, whose instinct compels them to live in society and obey a chief, are most capable of improvement, so is it with the races of mankind. Whether we look at it as a cause or a consequence, the more civilized always have the most artificial governments.

Journal of Researches 1839, 142

Wherever the European has trod, death seems to pursue the aboriginal. We may look to the wide extent of the Americas, Polynesia, the Cape of Good Hope, and Australia, and we shall find the same result. . . . The varieties of man seem to act on each other; in the same way as different species of animals—the stronger always extirpating the weaker.

Journal of Researches 1839, 520

On the whole, as a place of punishment [New South Wales, Australia] the object is scarcely gained; as a real system of reform it has failed, as perhaps would every other plan: but as a means of making men outwardly honest, —of converting vagabonds most useless in one hemisphere into active citizens of another, and thus giving birth to a new and splendid country—a grand centre of civilization—it has succeeded to a degree perhaps unparalleled in history.

Journal of Researches 1839, 532

You have an immense, incalculable advantage in living in a country [Australia] in which your children are sure to get on if industrious. I assure you that, though I am a rich man, when I think of the future I very often ardently wish I was settled in one of our Colonies, for I have now four sons (seven children in all, and more coming), and what on earth to bring them up to I do not know. A young man may here slave for years in any profession and not make a penny. Many people think that Californian gold will half ruin all those who live on the interest of accumulated gold or capital, and if that does happen I will certainly emigrate.

Darwin to Syms Covington,
23 November 1850, DCP 1370

I have received in a Manchester Newspaper a rather a good squib, showing that I have proved "might is right", & therefore that Napoleon [Emperor Napoleon III of France] is right & every cheating Tradesman is also right.

Darwin to Charles Lyell,
4 May [1860], DCP 2782

Obscure as is the problem of the advance of civilisation, we can at least see that a nation which produced during a lengthened period the greatest number of highly intellectual, energetic, brave, patriotic, and benevolent men, would generally prevail over less favoured nations.

Descent 1871, vol. 1, 180

The primary or fundamental check to the continued increase of man is the difficulty of gaining subsistence and of living in comfort. We may infer that this is the case from what we see, for instance, in the United States, where subsistence is easy and there is plenty of room.

If such means were suddenly doubled in Great Britain, our number would be quickly doubled. With civilised nations the above primary check acts chiefly by restraining marriages. The greater death rate of infants in the poorest classes is also very important; as well as the greater mortality at all ages, and from various diseases, of the inhabitants of crowded and miserable houses. The effects of severe epidemics and wars are soon counterbalanced, and more than counterbalanced, in nations placed under favourable conditions. Emigration also comes in aid as a temporary check, but not to any great extent with the extremely poor classes.

Descent 1871, vol. 1, 131–32

With savages, the weak in body or mind are soon eliminated; and those that survive commonly exhibit a vigorous state of health. We civilised men, on the other hand, do our utmost to check the process of elimination; we build asylums for the imbecile, the maimed, and the sick; we institute poor-laws; and our medical men exert their utmost skill to save the life of every one to the last moment. There is reason to believe that vaccination has preserved thousands, who from a weak constitution would formerly have succumbed to small-pox. Thus the weak members of civilised societies propagate their kind. No one who has attended to the breeding of domestic animals will doubt that this must be highly injurious to the race of man.

Descent 1871, vol. 1, 168

In all civilised countries man accumulates property and bequeaths it to his children. So that the children in the same country do not by any means start fair in the race for success. But this is far from an unmixed evil; for without the accumulation of capital the arts could not progress; and it is chiefly through their power that the civilised races have extended, and are now everywhere extending, their range, so as to take the place of the lower races.

Descent 1871, vol. 1, 169

The presence of a body of well-instructed men, who have not to labour for their daily bread, is important to a degree which cannot be overestimated; as all high intellectual work is carried on by them, and on such work material progress of all kinds mainly depends, not to mention other and higher advantages. No doubt wealth when very great tends to convert men into useless drones, but their number is never large; and some degree of elimination here occurs, as we daily see rich men, who happen to be fools or profligate, squandering away all their wealth.

Descent 1871, vol. 1, 169–70

The men who are rich through primogeniture are able to select generation after generation the more beautiful and charming women; and these must generally be healthy in body and active in mind. The evil consequences, such as they may be, of the continued preservation of the same line of descent, without any selection, are checked by men of rank always wishing to increase their wealth and power; and this they effect by marrying heiresses.

Descent 1871, vol. 1, 170

There is apparently much truth in the belief that the wonderful progress of the United States, as well as the character of the people, are the results of natural selection; the more energetic, restless, and courageous men from all parts of Europe having emigrated during the last ten or twelve generations to that great country, and having there succeeded best.

Descent 1871, vol. 1, 179

To believe that man was aboriginally civilised and then suffered utter degradation in so many regions, is to take a pitiably low view of human nature. It is apparently a truer and more cheerful view that progress has been much more general than retrogression; that man has risen, though by slow and interrupted steps, from a lowly condition to the highest standard as yet attained by him in knowledge, morals, and religion.

Descent 1871, vol. 1, 183–84

The advancement of the welfare of mankind is a most intricate problem: all ought to refrain from marriage who cannot avoid abject poverty for their children; for poverty is not only a great evil, but tends to its own increase by leading to recklessness in marriage. On the other hand, as Mr. Galton has remarked, if the prudent avoid marriage, whilst the reckless marry, the inferior members will tend to supplant the better members of society.

Descent 1871, vol. 2, 403

With highly civilised nations continued progress depends in a subordinate degree on natural selection; for such nations do not supplant and exterminate one another as do savage tribes. Nevertheless the more intelligent members within the same community will succeed better in the long run than the inferior, and leave a more numerous progeny, and this is a form of natural selection. The more efficient causes of progress seem to consist of a good education during youth whilst the brain is impressible, and of a high standard of excellence, inculcated by the ablest and best men, embodied in the laws, customs and traditions of the nation, and enforced by public opinion.

Descent 1874, 143

PART 5

On Himself

Darwin, photograph by Julia Margaret Cameron, 1868. Reproduced with permission from Wellcome Library, London.

Religious Belief

Whilst on board the *Beagle* I was quite orthodox, and I remember being heartily laughed at by several of the officers (though themselves orthodox) for quoting the Bible as an unanswerable authority on some point of morality. I suppose it was the novelty of the argument that amused them.

Autobiography, 85

Among the scenes which are deeply impressed on my mind, none exceed in sublimity the primeval forests undefaced by the hand of man; whether those of Brazil, where the powers of Life are predominant, or those of Tierra del Fuego, where death and Decay prevail. Both are temples filled with the varied productions of the God of Nature: no one can stand in these solitudes unmoved, and not feel that there is more in man than the mere breath of his body.

Journal of Researches 1839, 604–5

In my Journal I wrote that whilst standing in the midst of the grandeur of a Brazilian forest, "it is not possible to give an adequate idea of the higher feelings of wonder, admiration, and devotion which fill and elevate the mind." I well remember my conviction that there is more in man than the mere breath of his body. But now the grandest scenes would not cause any such convictions and feelings to rise in my mind.

Autobiography, 91

I gradually came to disbelieve in Christianity as a divine revelation. . . . But I was very unwilling to give up my belief;—I feel sure of this for I can well remember often and often inventing day-dreams of old letters between distinguished Romans and manuscripts being discovered at Pompeii or elsewhere which confirmed in the most striking manner all that was written in the Gospels. But I found it more and more difficult, with free scope given to my imagination, to invent evidence which would suffice to convince me. Thus disbelief crept over me at a very slow rate, but was at last complete. The rate was so slow that I felt no distress, and have never since doubted even for a single second that my conclusion was correct.

Autobiography, 86–87

I can indeed hardly see how anyone ought to wish Christianity to be true; for if so the plain language of the text seems to show that the men who do not believe, and this would include my Father, Brother and almost all my best friends, will be everlastingly punished. And this is a damnable doctrine.

Autobiography, 87

That there is much suffering in the world no one disputes. Some have attempted to explain this in reference to man by imagining that it serves for his moral improvement. But the number of men in the world is as nothing compared with that of all other sentient beings, and these often suffer greatly without any moral improvement. A being so powerful and so full of knowledge as a God who could create the universe, is to our finite minds omnipotent and omniscient, and it revolts our understanding to suppose that his benevolence is not unbounded, for what advantage can there be in the sufferings of millions of the lower animals throughout almost endless time? This very old argument from the existence of suffering against the existence of an intelligent first cause seems to me a strong one; whereas, as just remarked, the presence of much suffering agrees well with the view that

all organic beings have been developed through variation and natural selection.
Autobiography, 90

There is no evidence that man was aboriginally endowed with the ennobling belief in the existence of an Omnipotent God. On the contrary there is ample evidence, derived not from hasty travellers, but from men who have long resided with savages, that numerous races have existed and still exist, who have no idea of one or more gods, and who have no words in their languages to express such an idea. The question is of course wholly distinct from that higher one, whether there exists a Creator and Ruler of the universe; and this has been answered in the affirmative by the highest intellects that have ever lived.
Descent 1871, vol. 1, 65

As soon as the important faculties of the imagination, wonder, and curiosity, together with some power of reasoning, had become partially developed, man would naturally have craved to understand what was passing around him, and have vaguely speculated on his own existence.
Descent 1871, vol. 1, 65

The feeling of religious devotion is a highly complex one, consisting of love, complete submission to an exalted and mysterious superior, a strong sense of dependence, fear, reverence, gratitude, hope for the future, and perhaps other elements. No being could experience so complex an emotion until advanced in his intellectual and moral faculties to at least a moderately high level. Nevertheless we see some distant approach to this state of mind, in the deep love of a dog for his master, associated with complete submission, some fear, and perhaps other feelings.

Descent 1871, vol. 1, 68

With respect to immortality, nothing shows me how strong and almost instinctive a belief it is, as the consideration of the view now held by most physicists, namely that the sun with all the planets will in time grow too cold for life, unless indeed some great body dashes into the sun and thus gives it fresh life.—Believing as I do that man in the distant future will be a far more perfect creature than he now is, it is an intolerable thought that he and all other sentient beings are doomed to complete annihilation after such long-continued slow progress. To those who fully admit the immortality of the human soul, the destruction of our world will not appear so dreadful.

Autobiography, 92

When thus reflecting I feel compelled to look to a First Cause having an intelligent mind in some degree analogous to that of man; and I deserve to be called a Theist. This conclusion was strong in my mind about the time, as far as I can remember, when I wrote the *Origin of Species*; and it is since that time that it has very gradually with many fluctuations become weaker.

Autobiography, 92–93

The state of mind which grand scenes formerly excited in me, and which was intimately connected with a belief in God, did not essentially differ from that which is often called the sense of sublimity; and however difficult it may be to explain the genesis of this sense, it can hardly be advanced as an argument for the existence of God, any more than the powerful though vague and similar feelings excited by music.

Autobiography, 91–92

Mr Darwin begs me to say that he receives so many letters that he cannot answer them all.

He considers that the theory of evolution is quite compatible with the belief in a God; but that you must remember that different persons have different definitions of what they mean by God.

> Darwin to N. A. von Mengden,
> 8 April 1879, in the hand of
> Emma Darwin, DCP 11981

What my own views may be is a question of no consequence to any one but myself. But as you ask, I may state that my judgment often fluctuates. . . . In my most extreme fluctuations I have never been an atheist in the sense of denying the existence of a God.—I think that generally (& more and more so as I grow older), but not always, that an agnostic would be the most correct description of my state of mind.

> Darwin to John Fordyce,
> 7 May 1879, DCP 12041

Though I am a strong advocate for free thought on all subjects, yet it appears to me (whether rightly or wrongly) that direct arguments against Christianity & theism produce hardly any effect on the public; & freedom of thought is best promoted by the gradual illumination of men's minds, which follows from the advance of science. It has, therefore, been always my object to avoid writing on religion, & I have confined myself to science. I may, however, have been unduly biassed by the pain which it would give some members of my family, if I aided in any way direct attacks on religion.

Darwin to E. B. Aveling,
13 October 1880, DCP 12757

Dear Sir, I am sorry to have to inform you that I do not believe in the Bible as a divine revelation & therefore not in Jesus Christ as the son of God.

Darwin to Frederick McDermott,
24 November 1880, Bonhams sale
catalogue, New York, September 2015

Then the talk fell upon Christianity, and these remarkable words were uttered: "I never gave up Christianity until I was forty years of age." . . . On further inquiry, he told us that he had, when of mature years, investigated the claims of Christianity. Asked why he had abandoned it, the reply, simple and all-sufficient, was: "It is not supported by evidence."

Aveling 1883, 5–6, 7

There is one sentence in the Autobiography which I very much wish to omit, no doubt partly because your father's opinion that *all* morality has grown up by evolution is painful to me; but also because where this sentence comes in, it gives one a sort of shock—and would give an opening to say, however unjustly, that he considered all spiritual beliefs no higher than hereditary aversions or likings, such as the fear of monkeys towards snakes. . . . I should wish if possible to avoid giving pain to your father's religious friends who are deeply attached to him, and I picture to myself the way that sentence would strike them, even those so liberal as Ellen Tollett and Laura, much more Admiral Sullivan, Aunt Caroline, &c., and even the old servants.

> Emma Darwin to Francis Darwin,
> quoted in Barlow 1958, 93, note

Health

On Oct. 24th [1831], I took up my residence at Plymouth, and remained there until December 27th when the *Beagle* finally left the shores of England for her circumnavigation of the world. We made two earlier attempts to sail, but were driven back each time by heavy gales. These two months at Plymouth were the most miserable which I ever spent, though I exerted myself in various ways. I was out of spirits at the thought of leaving all my family and friends for so long a time, and the weather seemed to me inexpressibly gloomy. I was also troubled with palpitations and pain about the heart, and like many a young ignorant man, especially one with a smattering of medical knowledge, was convinced that I had heart-disease. I did not consult any doctor, as I fully expected to hear the verdict that I was not fit for the voyage, and I was resolved to go at all hazards.

Autobiography, 79–80

All last autumn and winter my health grew worse and worse; incessant sickness, tremulous hands and swimming head; I thought I was going the way of all flesh. Having heard of much success in some cases from the Cold Water Cure, I determined to give up all attempts to do anything and come here [Malvern] and put myself under Dr. [William] Gully. It has answered to a considerable extent: my sickness much checked and considerable strength gained. Dr. G., moreover, (and I hear he rarely speaks confidently) tells me he has little doubt but that he can cure me, in the course of time. Time however it will take.

Darwin to J. S. Henslow,
6 May 1849, DCP 1241

I am very doubtful whether I shall be up for [the Philosophical] Club; owing to Boys holidays drawing to a close, & sickness in our house. My wife often ails, & Lenny has very frequent bad days with badly intermittent pulse.—We escaped a considerable anxiety in George having apparently a regular low fever, but it died away & has spoiled only a fortnight of his holidays. Oh health, health, you are my daily & nightly bug-bear & stop all enjoyment in life. Etty keeps very weak.—But I really beg pardon, it is very foolish & weak to howl this way. Everyone has got his heavy burthen in this world.

> Darwin to J. D. Hooker,
> 15 January [1858], DCP 2203

I have lately spent a very pleasant week at Moor Park, & Hydropathy & idleness did me wonderful good & I walked one day 4 ½ miles,—a quite Herculean feat for me!

> Darwin to W. D. Fox,
> 13 November [1858], DCP 2360

My health has been very bad; & I am becoming as weak as a child, & incapable of doing anything whatever except my 3 hours daily work at Proof-sheets.—God knows whether I shall ever be good for anything again—perhaps a long rest & hydropathy may do something.

> Darwin to J. D. Hooker,
> 1 September [1859], DCP 2485

I have been talking with my wife & she joins heartily in asking whether Mrs. Huxley would not come here for a fortnight & bring all the children & nurse. But I must make it clear that this House is dreadfully dull & melancholy. My wife lives upstairs with my girl & she would see little of Mrs. Huxley, except at meal times, & my stomach is so habitually bad that I never spend the whole evening even with our nearest relations. If Mrs Huxley could be induced to come, she must look at this house, just as if it were a country inn, to which she went for a change of air.

Darwin to T. H. Huxley,
22 February [1861], DCP 3066

If [J. S.] Henslow, you thought, would really like to see me, I would of course start at once. The thought had [at] once occurred to me to offer, & the sole reason why I did not was that the journey with the agitation would cause me probably to arrive utterly prostrated. I sh^d. be certain to have severe vomiting afterwards, but that would not much signify, but I doubt whether I could stand the agitation at the time. I never felt my weakness a greater evil.... I suppose there is some Inn at which I could stay, for I sh^d not like to be in the House (even if you could hold me) as my retching is apt to be extremely loud.

Darwin to J. D. Hooker,
23 [April 1861], DCP 3125

I by no means thought that I produced a "tremendous effect" on Linn. Soc [the Linnean

Society of London]; but by Jove the Linn. Soc. produced a tremendous effect on me for I vomited all night & could not get out of bed till late next evening, so that I just crawled home.—I fear I must give up trying to read any paper or speak. It is a horrid bore I can do nothing like other people.

Darwin to J. D. Hooker,
9 [April 1862], DCP 3500

Hurrah! I have been 52 hours without vomiting!!

Darwin to J. D. Hooker,
26[–27] March [1864], DCP 4436

My sickness is not from mere irritability of stomach but is always caused by acid & morbid secretions. . . . For 25 years extreme spasmodic daily & nightly flatulence: occasional vomiting; on two occasions prolonged during months. Extreme secretion of saliva with flatulence. Vomiting preceded by shivering, hysterical crying dying sensations or half-faint. & copious very pallid urine. Now vomiting & every paroxysm of flatulence preceded by singing of ears, rocking, treading on air & vision. focus & black dots All fatigues, specially reading, brings on these Head symptoms. . . . (What I vomit intensely acid, slimy (sometimes bitter) corrodes teeth.) Doctors puzzled, say suppressed gout Family gouty. . . . Feet coldish.–Pulse 58 to 62–or slower & like thread. Appetite good—not thin. Evacuation regular & good. Urine scanty (because do not drink) often much pinkish sediment when cold—seldom headach or nausea.—Cannot walk above ½ mile—always tired—conversation or excitement tires me most. . . . Eczema– (now constant) lumbago–fundament–rash.

Darwin to Dr. John Chapman,
16 May [1865], DCP 4834

Politics

I never knew the newspapers so profoundly interesting [about the American Civil War, 1861–65]. N. America does not do England justice: I have not seen or heard of a soul who is not with the North. Some few, & I am one, even wish to God, though at the loss of millions of lives, that the North would proclaim a crusade against Slavery. In the long run, a million horrid deaths would be amply repaid in the cause of humanity. . . . Great God how I sh^d like to see that greatest curse on Earth, Slavery, abolished.

Darwin to Asa Gray,
5 June [1861], DCP 3176

It is surprising to me that you sh^d. have strength of mind to care for science, amidst the awful events daily occurring in your country. I daily look at the Times with almost as much interest as an American could do. When will peace come: it is dreadful to think of the desolation of large parts of your magnificent country; & all the speechless misery suffered by many.

> Darwin to Asa Gray,
> 10–20 June [1862], DCP 3595

Slavery draws me one day one way & another day another way. But certainly the Yankees are utterly detestable towards us.—What a new idea of Struggle for existence being necessary to try & purge a government! I daresay it is very true.

> Darwin to J. D. Hooker,
> 24 December [1862], DCP 3875

Our aristocracy is handsomer (more hideous according to a Chinese or Negro) than middle classes from pick of women; but oh what a scheme is primogeniture for destroying N. Selection.

> Darwin to A. R. Wallace,
> 28 [May 1864], DCP 4510

The great sin of Slavery has been almost universal, and slaves have often been treated in an infamous manner. As barbarians do not regard the opinion of their women, wives are commonly treated like slaves. Most savages are utterly indifferent to the sufferings of strangers, or even delight in witnessing them.

> *Descent* 1871, vol. 1, 94

Science

It appears to me, the doing what *little* one can to encrease the general stock of knowledge is as respectable an object of life, as one can in any likelihood pursue.

> Darwin to E. C. Darwin,
> 22 May–14 July 1833, CDP 206

During these two years [1837–39] I took several short excursions as a relaxation, and one longer one to the parallel roads of Glen Roy [Scotland], an account of which was published in the *Philosophical Transactions [of the Royal Society]*. This paper was a great failure, and I am ashamed of it. Having been deeply impressed with what I had seen of the elevation of the land in S. America, I attributed the parallel lines to the action of the sea; but I had to give up this view when [Louis] Agassiz propounded his glacier-lake theory. Because no other explanation was possible under our then state of knowledge, I argued in favour of sea-action; and my error has been a good lesson to me never to trust in science to the principle of exclusion.

Autobiography, 84

I am a firm believer that without speculation there is no good & original observation.

Darwin to A. R. Wallace,
22 December 1857, DCP 2192

How odd it is that every one should not see that all observation must be for or against some view, if it is to be of any service.

Darwin to Henry Fawcett,
18 September [1861], DCP 3257

I am sometimes half tempted to give up Species & stick to experiments; they are much better fun.

Darwin to J. D. Hooker,
9 February [1862], DCP 3440

I did not fully appreciate your insect-diving-case before your last note; nor had I any idea that the fact was new, though new to me. It is really very interesting. Of course you will publish an account of it. You will then say whether the insect can fly well through the air. My wife asked how did he find out that it stayed 4 hours under water without breathing; I answered at

once "M^rs. Lubbock sat four hours watching".
I wonder whether I am right.

> Darwin to John Lubbock,
> 5 September [1862], DCP 3713

I am like a gambler, & love a wild experiment.

> Darwin to J. D. Hooker,
> 26 [March 1863], DCP 4061

Forgive me for suggesting one caution; as Demosthenes said, "action, action, action" was the soul of eloquence, so is caution almost the soul of science. Pray bear in mind that if a naturalist is once considered, though unjustly, as not quite trust worthy, it takes long years before he can recover his reputation for accuracy.

> Darwin to Anton Dohrn,
> 4 January 1870, DCP 7070

It has often and confidently been asserted, that man's origin can never be known: but ignorance more frequently begets confidence than does knowledge: it is those who know little, and not those who know much, who so positively assert that this or that problem will never be solved by science.

> *Descent* 1871, vol. 1, 3

False facts are highly injurious to the progress of science, for they often long endure; but false views, if supported by some evidence, do little harm, as every one takes a salutary pleasure in proving their falseness; and when this is done, one path towards error is closed and the road to truth is often at the same time opened.
Descent 1871, vol. 2, 385

I have been speculating last night what makes a man a discoverer of undiscovered things, & a most perplexing problem it is.—Many men who are very clever—much cleverer than discoverers—never originate anything. As far as I can conjecture, the art consists in habitually searching for the causes and meaning of everything which occurs. This implies sharp observation & requires as much knowledge as possible of the subject investigated.
Darwin to Horace Darwin,
[15 December 1871], DCP 8107

We had grand fun one afternoon, for George [Darwin] hired a medium, who made the chairs, a flute, a bell & candlestick & fiery points jump about in my Brother's dining room, in a manner that astounded everyone & took away all their breaths. It was in the dark, but George & Hensleigh [Wedgwood] held the medium's hands & feet on both sides all the time. I found it so hot & tiring that I went away before all these astounding miracles or jugglery took place. How the man could possibly do what was done, passes my understanding. I came down stairs, & saw all the chairs, etc. etc. on the table which had been lifted over the heads of those sitting around it. The Lord have mercy on us all, if we have to believe in such rubbish. F. Galton was there and says it was a good séance.

> Darwin to J. D. Hooker,
> 18 January [1874], DCP 9247

Perhaps you saw in the papers that the Turin Soc. honoured me to an extraordinary degree by awarding me the Bressa prize. Now it occurred to me that if your Station [Stazione Zoologica di Napoli] wanted some piece of apparatus of about the value of 100 £ I shd. very much like to be allowed to pay for it.

> Darwin to Anton Dohrn, 15 February
> 1880, quoted in Gröben 1982, 70

From quotations which I had seen I had a high notion of Aristotle's merits, but I had not the most remote notion what a wonderful man he was. Linnaeus and Cuvier have been my two gods, though in very different ways, but they were mere school-boys to old Aristotle.

> Darwin to William Ogle, 22 February 1882, quoted in Gotthelf 1999, 4

I have steadily endeavoured to keep my mind free, so as to give up any hypothesis, however much beloved (and I cannot resist forming one on every subject), as soon as facts are shown to be opposed to it. Indeed I have had no choice but to act in this manner, for with the exception of the Coral Reefs, I cannot remember a single first-formed hypothesis which had not after a time to be given up or greatly modified. This has naturally led me to distrust greatly deductive reasoning in the mixed sciences.

> *Autobiography*, 141

Writing

I am just now beginning to discover the difficulty of expressing one's ideas on paper. As long as it consists solely of description it is pretty easy; but where reasoning comes into play, to make a proper connection, a clearness & a moderate fluency, is to me, as I have said, a difficulty of which I had no idea.

> Darwin to C. S. Darwin,
> 29 April 1836, DCP 301

I find the style [of *On the Origin of Species*] incredibly bad, & most difficult to make clear & smooth.

> Darwin to John Murray,
> 14 June [1859], DCP 2469

To me, observing is much better sport than writing.

> Darwin to Henry Fawcett,
> 18 September [1861], DCP 3257

In writing, he sometimes showed the same tendency to strong expressions as he did in conversation. Thus in the *Origin*, p. 440, there is a description of a larval cirripede, "with six pairs of beautifully constructed natatory legs, a pair of magnificent compound eyes, and extremely complex antennæ." We used to laugh at him for this sentence, which we compared to an advertisement. This tendency to give himself up to the enthusiastic turn of his thought, without fear of being ludicrous, appears elsewhere in his writings.

> F. Darwin in *Life and Letters*, vol. 1, 156

Write the book carefully and then go over it again, crossing out every sentence that looks like particularly fine composition.

> Advice to H. W. Bates, quoted in
> Obituary of Henry Walter Bates,
> *Proceedings of the Royal
> Geographical Society* 14 (4): 251

There seems to be a sort of fatality in my mind leading me to put at first my statement and proposition in a wrong or awkward form. Formerly I used to think about my sentences before writing them down; but for several years I have found that it saves time to scribble in a vile hand whole pages as quickly as I possibly can, contracting half the words; and then correct deliberately. Sentences thus scribbled down are often better ones than I could have written deliberately.

Autobiography, 136–37

I may mention that I keep from thirty to forty large portfolios, in cabinets with labelled shelves, into which I can at once put a detached reference or memorandum. I have bought many books and at their ends I make an index of all the facts that concern my work; or, if the book is not my own, write out a separate abstract, and of such abstracts I have a large drawer full. Before beginning on any subject I look to all the short indexes and make a general and classified index, and by taking the one or more proper portfolios I have all the information collected during my life ready for use.

Autobiography, 137–38

Please read the Ch. [proof sheets of *Descent of Man*] first right through without a pencil in your hand, that you may judge of general scheme; as, also, I particularly wish to know whether parts are extra tedious; but remember that M.S is always much more tedious than print. . . . I fear parts are too like a Sermon: who wd ever have thought that I shd. turn parson?

Darwin to Henrietta Darwin,
[8 February 1870], DCP 7124

I have worked through (and it is hard work), half of the 2nd chapter on mind [proofs of *Descent of Man*], and your corrections and suggestions are excellent. I have adopted the greater number, and I am sure that they are very great improvements. Some of the transpositions are most just. You have done me real service; but, by Jove, how hard you must have worked, and how thoroughly you have mastered my MS. I am pleased with this chapter now that it comes fresh to me. Your affectionate, and admiring and obedient father, C. D.

Darwin to Henrietta Darwin,
[March] 1870, *Emma Darwin*, vol. 2, 230

Dogs

My father was always fond of dogs, and as a young man had the power of stealing away the affections of his sisters' pets; at Cambridge, he won the love of his cousin W.D. Fox's dog, and this may perhaps have been the little beast which used to creep down inside his bed and sleep at the foot every night. My father had a surly dog, who was devoted to him, but unfriendly to every one else, and when he came back from the Beagle voyage, the dog remembered him, but in a curious way, which my father was fond of telling. He went into the yard and shouted in his old manner; the dog rushed out and set off with him on his walk, showing no more emotion or excitement than if the same thing had happened the day before, instead of five years ago.

           F. Darwin, *Life and Letters*, vol. 1, 113

The dog most closely associated with my father was . . . Polly, a rough, white fox-terrier. . . . My father used to make her catch biscuits off her nose, and had an affectionate and mock-solemn way of explaining to her beforehand that she must "be a very good girl." She had a mark on her back where she had been burnt, and where the hair had re-grown red instead of white, and my father used to commend her for this tuft of hair as being in accordance with his theory of pangenesis; her father had been a red bull-terrier, thus the red hair appearing after the burn showed the presence of latent red gemmules.

F. Darwin, *Life and Letters*, vol. 1, 113–14

I formerly possessed a large dog, who, like every other dog, was much pleased to go out walking. He showed his pleasure by trotting gravely before me with high steps, head much raised, moderately erected ears, and tail carried aloft but not stiffly. Not far from my house a path branches off to the right, leading to the hot-house, which I used often to visit for a few moments, to look at my experimental plants. This was always a great disappointment to the dog, as he did not know whether I should continue my walk; and the instantaneous and complete change of expression which came over him, as soon as my body swerved in the least towards the path (and I sometimes tried this as an experiment) was laughable. His look of dejection was known to every member of the family, and was called his *hot-house face*. This consisted in the head drooping much, the whole body sinking a little and remaining motionless; the ears and tail falling suddenly down, but the tail was by no means wagged. With the falling of the ears and of his great chaps, the eyes became much changed in appearance, and I fancied that they looked less bright. His aspect was that of piteous, hopeless dejection; and it was, as I have said, laughable, as the cause was so slight.

Expression, 59–60

A female terrier of mine lately had her puppies destroyed, and though at all times a very affectionate creature, I was much struck with the manner in which she then tried to satisfy her instinctive maternal love by expending it on me; and her desire to lick my hands rose to an insatiable passion.

Expression, 120

The love of a dog for his master is notorious; in the agony of death he has been known to caress his master, and every one has heard of the dog suffering under vivisection, who licked the hand of the operator; this man, unless he had a heart of stone, must have felt remorse to the last hour of his life.

Descent 1871, vol. 1, 40

Anti-vivisection

I would gladly punish severely any one who operated on an animal not rendered insensible, if the experiment made this possible; but here again I do not see that a magistrate or jury could possibly determine such a point. Therefore I conclude, if (as is likely) some experiments have been tried too often, or anæsthetics have not been used when they could have been, the cure must be in the improvement of humanitarian feelings.

Darwin to Henrietta (Darwin)
Litchfield, 4 January1875,
Life and Letters, vol. 3, 202

I have all my life been a strong advocate for humanity to animals, and have done what I could in my writings to enforce this duty. Several years ago, when the agitation against physiologists commenced in England, it was asserted that inhumanity was here practised and useless suffering caused to animals; and I was led to think that it might be advisable to have an Act of Parliament on the subject. I then took an active part in trying to get a Bill passed, such as would have removed all just cause of complaint, and at the same time have left physiologists free to pursue their researches—a Bill very different from the Act which has since been passed.

Darwin to Frithiof Holmgren,
The Times, 18 April 1881, 10

Physiology cannot possibly progress except by means of experiments on living animals, and I feel the deepest conviction that he who retards the progress of physiology commits a crime against mankind. Any one who remembers, as I can, the state of this science half a century ago must admit that it has made immense progress, and it is now progressing at an ever-increasing rate.

Darwin to Frithiof Holmgren,
The Times, 18 April 1881, 10

Mr. Darwin eventually became the centre of an adoring clique of vivisectors who (as his Biography shows) plied him incessantly with encouragement to uphold their practice, till the deplorable spectacle was exhibited of a man who would not allow a fly to bite a pony's neck, standing forth before all Europe (in his celebrated letter to Prof. [Frithiof] Holmgren of Sweden) as the advocate of Vivisection.

Cobbe 1894, vol. 2, 128

Nature

I cannot tell you how I enjoyed some of these views [in the Cordilliera].—it is worth coming from England once to feel such intense delight. At an elevation from 10–12000 ft. there is a transparency in the air & a confusion of distances & a sort of stillness which gives the sensation of being in another world, & when to this is joined, the picture so plainly drawn of the great epochs of violence, it causes in the mind a most strange assemblage of ideas.

Darwin to J. S. Henslow,
18 April 1835, DCP 274

While sailing in these latitudes on one very dark night, the sea presented a wonderful and most beautiful spectacle. There was a fresh breeze, and every part of the surface, which during the day is seen as foam, now glowed with a pale light. The vessel drove before her bows two billows of liquid phosphorus, and in her wake she was followed by a milky train. As far as the eye reached, the crest of every wave was bright, and the sky above the horizon, from the reflected glare of these livid flames, was not so utterly obscure, as over the rest of the heavens.

Journal of Researches 1839, 190–91

At first, from the waterfalls and number of dead trees [in Tierra del Fuego], I could hardly crawl along; but the bed of the stream soon became a little more open, from the floods having swept the sides. I continued slowly to advance for an hour along the broken and rocky banks; and was amply repaid by the grandeur of the scene. The gloomy depth of the ravine well accorded with the universal signs of violence. On every side were lying irregular masses of rock and up-torn trees; other trees, though still erect, were decayed to the heart and ready to fall. The entangled mass of the thriving and the fallen reminded me of the forests within the tropics;—yet there was a difference; for in these still solitudes, Death, instead of Life, seemed the predominant spirit.

Journal of Researches 1839, 231

Magnificent glaciers extended from the mountain side to the water's edge. It is scarcely possible to imagine any thing more beautiful than the beryl-like blue of the glacier, and especially when contrasted with the dead white of an expanse of snow. As fragments fell from the glacier into the water, they floated away, and the channel with its icebergs represented in miniature the polar sea.

Journal of Researches 1839, 243–44

When we reached the crest [of the Portillo pass] and looked backwards, a glorious view was presented. The atmosphere resplendently clear; the sky an intense blue; the profound valleys; the wild broken forms; the heaps of ruins, piled up during the lapse of ages; the bright-coloured rocks, contrasted with the quiet mountains of snow; all these together produced a scene I never could have figured to my imagination. Neither plant nor bird, excepting a few condors wheeling around the higher pinnacles, distracted the attention from the inanimate mass. I felt glad I was alone: it was like watching a thunderstorm, or hearing a chorus of the Messiah in full orchestra.

Journal of Researches 1839, 394

The weather is quite delicious. Yesterday after writing to you I strolled a little beyond the glade for an hour & half & enjoyed myself—the fresh yet dark green of the grand Scotch Firs, the brown of the catkins of the old Birches with their white stems & a fringe of distant green from the larches, made an excessively pretty view.—At last I fell fast asleep on the grass & awoke with a chorus of birds singing around me, & squirrels running up the trees & some Woodpeckers laughing, & it was as pleasant a rural scene as ever I saw, & I did not care one penny how any of the beasts or birds had been formed.—

> Darwin to Emma Darwin,
> [28 April 1858], DCP 2261

I quite agree how humiliating the slow progress of man is; but everyone has his own pet horror, & this slow progress, or even personal annihilation sinks in my mind into insignificance compared with the idea, or rather I presume certainty, of the sun some day cooling & we all freezing. To think of the progress of millions of years, with every continent swarming with good & enlightened men all ending in this; & with probably no fresh start until this our own planetary system has been again converted into red-hot gas.—Sic transit gloria mundi, with a vengeance.

> Darwin to J. D. Hooker,
> 9 February [1865], DCP 4769

Autobiographical

My mother died in July 1817, when I was a little over eight years old, and it is odd that I can remember hardly anything about her except her death-bed, her black velvet gown, and her curiously constructed work-table. I believe that my forgetfulness is partly due to my sisters, owing to their great grief, never being able to speak about her or mention her name; and partly to her previous invalid state.

Autobiography, 22

My father's mind was not scientific, and he did not try to generalise his knowledge under general laws; yet he formed a theory for almost everything which occurred. I do not think that I gained much from him intellectually.

Autobiography, 42

To my deep mortification my father once said to me, "You care for nothing but shooting, dogs, and rat-catching, and you will be a disgrace to yourself and all your family." But my father, who was the kindest man I ever knew, and whose memory I love with all my heart, must have been angry and somewhat unjust when he used such words.

Autobiography, 28

As far as I can judge of myself I worked to the utmost during the voyage from the mere pleasure of investigation, and from my strong desire to add a few facts to the great mass of facts in natural science. But I was also ambitious to take a fair place among scientific men.

Autobiography, 80–81

About this time [1839] I took much delight in Wordsworth's and Coleridge's poetry, and can boast that I read the *Excursion* twice through. Formerly Milton's *Paradise Lost* had been my chief favourite, and in my excursions during the voyage of the *Beagle*, when I could take only a single small volume, I always chose Milton.

Autobiography, 85

Looking backwards, I can now perceive how my love for science gradually preponderated over every other taste. . . . I discovered, though unconsciously and insensibly, that the pleasure of observing and reasoning was a much higher one than that of skill and sport. The primeval instincts of the barbarian slowly yielded to the acquired tastes of the civilized man.

Autobiography, 78

I should be very glad to hear about yourself, Mrs Fitzroy & the children. My life goes on like Clockwork, and I am fixed on the spot where I shall end it; we have four children, who & my wife are all well. My health, also, has rather improved, but I am a different man in strength and energy to what I was in old days, when I was your "Fly-catcher", on board the Beagle.

Darwin to Robert FitzRoy,
1 October 1846, DCP 1002

You do me injustice when you think that I work for fame: I value it to a certain extent; but, if I know myself, I work from a sort of instinct to try to make out truth.

Darwin to W. D. Fox,
24 [March 1859], DCP 2436

I send a Photograph of myself with my Beard. Do I not look venerable?

Darwin to Asa Gray,
28 May [1864], DCP 4511

I like this photograph much better than any other which has been taken of me.
>
> Endorsement of photograph
> by Julia Margaret Cameron, 1868

As for myself I believe that I have acted rightly in steadily following and devoting my life to science. I feel no remorse from having committed any great sin, but have often and often regretted that I have not done more direct good to my fellow creatures. My sole and poor excuse is much ill-health and my mental constitution, which makes it extremely difficult for me to turn from one subject or occupation to another. I can imagine with high satisfaction giving up my whole time to philanthropy, but not a portion of it; though this would have been a far better line of conduct.
>
> *Autobiography*, 95

My handwriting same as Grandfather.
>
> *Notebook M*, 83e

I rejoice that I have avoided controversies, and this I owe to [Charles] Lyell, who many years ago, in reference to my geological works, strongly advised me never to get entangled in a controversy, as it rarely did any good and caused a miserable loss of time and temper.
>
> *Autobiography*, 126

Pray give our very kind remembrances to Mrs. Gray. I know that she likes to hear men boasting,—it refreshes them so much. Now the tally with my wife in backgammon stands thus: she, poor creature, has won only 2490 games, whilst I have won, hurrah, hurrah, 2795 games.

Darwin to Asa Gray,
28 January 1876, DCP 10370

In one respect my mind has changed during the last twenty or thirty years. Up to the age of thirty, or beyond it, poetry of many kinds, such as the works of Milton, Gray, Byron, Wordsworth, Coleridge, and Shelley, gave me great pleasure, and even as a schoolboy I took intense delight in Shakespeare, especially in the historical plays. I have also said that formerly pictures gave me considerable, and music very great delight. But now for many years I cannot endure to read a line of poetry: I have tried lately to read Shakespeare, and found it so intolerably dull that it nauseated me. I have also almost lost any taste for pictures or music.—Music generally sets me thinking too energetically on what I have been at work on, instead of giving me pleasure. I retain some taste for fine scenery, but it does not cause me the exquisite delight which it formerly did.

Autobiography, 138

My mind seems to have become a kind of machine for grinding general laws out of large collections of facts, but why this should have caused the atrophy of that part of the brain alone, on which the higher tastes depend, I cannot conceive.

Autobiography, 139

Special talents: None except for business, as evinced by keeping accounts, replies to correspondence, and investing money very well. Very methodical in all my habits.

Response to questionnaire,
Life and Letters, vol. 3, 179

Everybody whom I have seen and who has seen your picture of me is delighted with it. I shall be proud some day to see myself suspended at the Linnean Society.

Darwin to John Collier,
16 February 1882, *More Letters*, vol. 1, 398

For my own part I would as soon be descended from that heroic little monkey, who braved his dreaded enemy in order to save the life of his keeper; or from that old baboon, who, descending from the mountains, carried away in triumph his young comrade from a crowd of astonished dogs—as from a savage who delights to torture his enemies, offers up bloody sacrifices, practises infanticide without remorse, treats his wives like slaves, knows no decency, and is haunted by the grossest superstitions.

Descent 1871, vol. 2, 404–5

I give and bequeath to each of my friends Sir Joseph Dalton Hooker and Thomas Henry Huxley Esquire the legacy or sum of one thousand pounds sterling free of legacy duty as a slight memorial of my life long affection and respect for them.

Last will and testament of
Charles Robert Darwin,
quoted in *Darwin Online*

With such moderate abilities as I possess, it is truly surprising that thus I should have influenced to a considerable extent the beliefs of scientific men on some important points.

Autobiography, 145

PART 6

Friends and Family

Darwin, photograph by Elliott & Fry, c. 1881. Reproduced with permission from the National Portrait Gallery London.

Friends and Contemporaries

Louis Agassiz: What a set of men you have in Cambridge! Both our Universities put together cannot furnish the like. Why, there is Agassiz, —he counts for three.

E. C. Agassiz 1890, 666

Robert Brown: I called on him two or three times before the voyage of the Beagle, and on one occasion he asked me to look through a microscope and describe what I saw. This I did, and believe now that it was the marvellous currents of protoplasm in some vegetable cell. I then asked him what I had seen; but he answered me, who was then hardly more than a boy and on the point of leaving England for five years, "That is my little secret." I suppose that he was afraid that I might steal his discovery.

Autobiography, 103–4

Samuel Butler: In 1879, I had a translation of Dr. Ernst Krause's *Life of Erasmus Darwin* published, and I added a sketch of his character and habits from materials in my possession. . . . Owing to my having accidentally omitted to mention that Dr. Krause had enlarged and corrected his article in German before it was translated, Mr Samuel Butler abused me with almost insane virulence. How I offended him so bitterly, I have never been able to understand.

Autobiography, 134–35

Thomas Carlyle: His talk was very racy and interesting, just like his writings, but he sometimes went on too long on the same subject. I remember a funny dinner at my brother's, where, amongst a few others, were [Charles] Babbage and [Charles] Lyell, both of whom liked to talk. Carlyle, however, silenced every one by haranguing during the whole dinner on the advantages of silence. After dinner, Babbage, in his grimmest manner, thanked Carlyle for his very interesting Lecture on Silence. . . . He has been all-powerful in impressing some grand moral truths on the minds of men. On the other hand, his views about slavery were revolting. In his eyes might was right.

Autobiography, 112–13

Erasmus Darwin: Dr. Darwin has been frequently called an atheist, whereas in every one of his works distinct expressions may be found showing that he fully believed in God as the Creator of the universe.... Although Dr. Darwin was certainly a theist in the ordinary acceptation of the term, he disbelieved in any revelation. Nor did he feel much respect for Unitarianism, for he used to say that "Unitarianism was a feather-bed to catch a falling Christian."

C. Darwin 1879, 44–45

Erasmus Alvey Darwin: My brother Erasmus possessed a remarkably clear mind, with extensive and diversified tastes and knowledge in literature, art, and even in science. For a short time he collected and dried plants, and during a somewhat longer time experimented in chemistry. He was extremely agreeable, and his wit often reminded me of that in the letters and works of Charles Lamb. He was very kindhearted; but his health from his boyhood had been weak, and as a consequence he failed in energy.

Autobiography, 42

Robert FitzRoy: FitzRoy's character was a singular one, with many very noble features: he was devoted to his duty, generous to a fault, bold, determined, indomitably energetic, and an ardent friend to all under his sway. He would undertake any sort of trouble to assist those whom he thought deserved assistance.... His temper was usually worst in the early morning, and with his eagle eye he could generally detect something amiss about the ship, and was then unsparing in his blame. The junior officers when they relieved each other in the forenoon used to ask "whether much hot coffee had been served out this morning,—" which meant how was the Captain's temper?

Autobiography, 72–73

William Darwin Fox: I am sure, if a long voyage may have some injurious tendencies to a person's character, it has the one good one of teaching him to appreciate & dearly love his friends & relations.

Darwin to W. D. Fox,
15 February 1836, DCP 299

William Ewart Gladstone: What an honor that such a great man should come to visit me!

Morley 1903, vol. 2, 562

Asa Gray: My conclusion is that you have made a mistake in being a Botanist, you ought to have been a Lawyer, & you would have rolled in wealth by perverting the truth, instead of studying the living truths of this world.

> Darwin to Asa Gray,
> 22 July [1860], DCP 2876

Asa Gray: I said in a former letter that you were a Lawyer; but I made a gross mistake, I am sure that you are a poet. No by Jove I will tell you what you are, a hybrid, a complex cross of Lawyer, Poet, Naturalist, & Theologian!—Was there ever such a monster seen before?

> Darwin to Asa Gray,
> 10 September [1860], DCP 2910

Ernst Haeckel: I have seldom seen a more pleasant, cordial & frank man. He is now in Madeira where he is going to work chiefly on the Medusæ. His great work is now published & I have a copy, but the German is so difficult I can make out but little of it, & I fear it is too large a work to be translated.

> Darwin to J.F.T. Müller,
> [before 10 December 1866], DCP 5261

John Stevens Henslow: He was deeply religious, and so orthodox that he told me one day he should be grieved if a single word of the Thirty-nine Articles were altered. His moral qualities were in every way admirable. He was free from every tinge of vanity or other petty feeling; and I never saw a man who thought so little about himself or his own concerns.

Autobiography, 64–65

Joseph Dalton Hooker: I became very intimate with Hooker, who has been one of my best friends throughout life. He is a delightfully pleasant companion and most kind-hearted. One can see at once that he is honourable to the backbone. His intellect is very acute, and he has great power of generalisation. He is the most untirable worker that I have ever seen, and will sit the whole day working with the microscope, and be in the evening as fresh and pleasant as ever. He is in all ways very impulsive and somewhat peppery in temper; but the clouds pass away almost immediately. . . . I have known hardly any man more lovable than Hooker.

Autobiography, 105–6

Joseph Dalton Hooker: The sight of your handwriting always rejoices the very cockles of my heart.

Darwin to J. D. Hooker,
15 January [1861], DCP 3047

Joseph Dalton Hooker: I have got your Photograph over my chimney-piece & like it much; but you look down so sharp on me that I shall never be bold enough to wriggle myself out of any contradiction.

Darwin to J. D. Hooker,
25 December [1868], DCP 6512

Alexander von Humboldt: I once met at breakfast at Sir R. Murchison's house, the illustrious Humboldt, who honoured me by expressing a wish to see me. I was a little disappointed with the great man, but my anticipations probably were too high. I can remember nothing distinctly about our interview, except that Humboldt was very cheerful and talked much.

Autobiography, 107

Thomas Henry Huxley: His mind is as quick as a flash of lightning and as sharp as a razor. He is the best talker whom I have known. He never writes and never says anything flat. From his conversation no one would suppose that he could cut up his opponents in so trenchant a manner as he can do and does do.... He is a splendid man and has worked well for the good of mankind.

Autobiography, 106

Thomas Henry Huxley: I often think that my friends (& you far beyond others) have good cause to hate me, for having stirred up so much mud, & led them into so much odious trouble.—If I had been a friend of myself, I should have hated me. (how to make that sentence good English I know not.) But remember if I had not stirred up the mud some one else certainly soon would.

Darwin to T. H. Huxley,
3 July [1860], DCP 2854

John Lubbock: It made me grieve his taking to Politicks, & though I grieve that he has lost his Election, yet I suppose now that he is once bitten he will never give up Politicks, & Science is done for. Many men can make fair M.P.s, & how few can work in science like him.

Darwin to J. D. Hooker,
[29 July 1865], DCP 4874

Charles Lyell: I saw more of Lyell than of any other man both before and after my marriage.

His mind was characterised, as it appeared to me, by clearness, caution, sound judgment and a good deal of originality.... His delight in science was ardent, and he felt the keenest interest in the future progress of mankind. He was very kind-hearted, and thoroughly liberal in his religious beliefs or rather disbeliefs; but he was a strong theist. His candour was highly remarkable. He exhibited this by becoming a convert to the Descent theory, though he had gained much fame by opposing Lamarck's views, and this after he had grown old. He reminded me that I had many years before said to him, when discussing the opposition of the old school of geologists to his new views, "What a good thing it would be if every scientific man was to die when sixty years old, as afterwards he would be sure to oppose all new doctrines." But he hoped that now he might be allowed to live.

Autobiography, 100–101

James Mackintosh: One of my autumnal visits to Maer in 1827 was memorable from meeting there Sir J. Mackintosh, who was the best converser I ever listened to. I heard afterwards with a glow of pride that he had said, "There is something in that young man that interests me." This must have been chiefly due to his perceiving that I listened with much interest to everything which he said, for I was as ignorant as a pig about his subjects of history, politicks and moral philosophy. To hear of praise from an eminent person, though no doubt apt or certain to excite vanity, is, I think, good for a young man, as it helps to keep him in the right course.

Autobiography, 55

Richard Owen: It is painful to be hated in the intense degree with which Owen hates me.
Darwin to Charles Lyell,
10 April [1860], DCP 2754

John Ruskin: It was very acute of Mr Ruskin to know that I have a deep & tender interest about the brightly coloured hinder half of certain monkeys.
Darwin to Victor A.E.G. Marshall,
7 [September] 1879, quoted in
Healey 2001, 306

Herbert Spencer: Herbert Spencer's conversation seemed to me very interesting, but I did not like him particularly, and did not feel that I could easily have become intimate with him. I think that he was extremely egotistical. After reading any of his books, I generally feel enthusiastic admiration for his transcendent talents, and have often wondered whether in the distant future he would rank with such great men as Descartes, Leibnitz, etc., about whom, however, I know very little. Nevertheless I am not conscious of having profited in my own work by Spencer's writings. His deductive manner of treating every subject is wholly opposed to my frame of mind. His conclusions never convince me: and over and over again I have said to myself, after reading one of his discussions, "Here would be a fine subject for half-a-dozen years' work."

Autobiography, 108–9

Alfred Russel Wallace: Your modesty and candour are very far from new to me. I hope it is a satisfaction to you to reflect,—& very few things in my life have been more satisfactory to me—that we have never felt any jealousy towards each other, though in one sense rivals. I believe that I can say this of myself with truth, & I am absolutely sure that it is true of you.

Darwin to A. R. Wallace,
20 April [1870], DCP 7167

Josiah Wedgwood II: He was the very type of an upright man with the clearest judgment. I do not believe that any power on earth could have made him swerve an inch from what he considered the right course. I used to apply to him in my mind, the well-known ode of Horace, now forgotten by me, in which the words "nec vultus tyranni, &c.," come in ["The man who is tenacious of purpose in a rightful cause is not shaken by the frenzy of his fellow citizens clamoring for what is wrong or by the tyrant's threatening countenance." Horace Book III, ode iii].

Autobiography, 56

Reflections by His Contemporaries

William Allingham: Tall, yellow, sickly, very quiet. He has his meals at his own times, sees people or not as he chooses, has invalid's privileges in full, a great help to a studious man.

Allingham 1907, 184

Samuel Butler: It is doubtless a common practice for writers [Darwin] to take an opportunity of revising their works, but it is not common when a covert condemnation of an opponent [Butler] has been interpolated into a revised edition, the revision of which has been concealed, to declare with every circumstance of distinctness that the condemnation was written prior to the book which might appear to have called it forth, and thus lead readers to suppose that it must be an unbiassed opinion.

S. Butler, letter criticising
Darwin's *Life of Erasmus Darwin*,
31 January 1880, *Athenæum*, 155

Moncure Daniel Conway: This formidable man, speaking from the shelter of the English throne and from under the wings of the English Church itself, did not mean to give Dogmatic Christianity its deathblow; he meant to utter a simple theory of nature.

Conway 1904, 250

Anton Dohrn: I must confess, Darwin's personal appearance surprised me very much. I had expected to find a sick-looking man; instead I saw before me a tall, strong, grey bearded stature, full of life and cheerfulness and heart-winning amiability.

Gröben 1982, 93

Hugo De Vries: He has deep set eyes and in addition very protruding eyebrows, much more than one would say from his portrait. He is tall and thin and has thin hands, he walks slowly and uses a cane and has to stop from time to time. He is very much afraid of drafts and generally has to be very careful with his health. His speech is very lively, merry and cordial, not too quick and very clear.

Pas 1970, 187

Hugh Falconer: I am of opinion that Mr. Darwin is not only one of the most eminent naturalists of his day, but that hereafter he will be regarded as one of the Great Naturalists of all Countries and of all time.

> Nomination for the Copley
> Medal of the Royal Society,
> 25 October 1864, DCP 4644

Rev. George Ffinden [vicar of Down parish]: I confess that, perhaps, I am a bit sour over Darwin and his works. You see, I'm a Churchman first and foremost. He never came to church, and it was such a bad business for the parish, a bad example. He was, however, most amiable and benevolent and courteous, and very liberal. I remember his giving me a subscription for the church and the house restoration or building. "Of course," he told me, "I don't believe in this at all." "I don't suppose you do," I said to him. Quite candid on both sides.

> A visit to Darwin's village,
> *Evening News*, 12 February 1909, 4

William Darwin Fox: I suppose your destiny is to let your Brain destroy your Body.

> W. D. Fox to Darwin,
> 28 November [1864], DCP 4683

Francis Galton: I felt little difficulty in connecting with the *Origin of Species*, but devoured its contents and assimilated them as fast as they were devoured, a fact which perhaps may be ascribed to an hereditary bent of mind that both its illustrious author and myself have inherited from our common grandfather, Dr. Erasmus Darwin. I made occasional excursions to visit Charles Darwin at Down, usually at luncheon-time, always with a sense of the utmost veneration as well as of the warmest affection, which his invariably hearty greeting greatly encouraged. I think his intellectual characteristic that struck me most forcibly was the aptness of his questionings; he got thereby very quickly to the bottom of what was in the mind of the person he conversed with, and to the value of it.

Galton 1909, 288

Ernst Haeckel: There stepped out to meet me from the shady porch, overgrown with creeping plants, the great naturalist himself, a tall and venerable figure with the broad shoulders of an Atlas supporting a world of thoughts, his Jupiter-like forehead highly and broadly arched, as in the case of Goethe, and deeply furrowed by the plough of mental labour; his kindly, mild eyes looking forth under the shadow of prominent brows; his amiable mouth surrounded by a copious silver-white beard.

Haeckel 1882, 6

J. D. Hooker: Glorified friend! Your photograph tells me where [John Rogers] Herbert got his Moses for the Fresco in the House of Lords. —horns & halo & all.

J. D. Hooker to Darwin,
[11 June 1864], DCP 4529

Alexander von Humboldt: You have an excellent future ahead of you.

A. von Humboldt to Darwin,
18 September 1839, DCP 534

T. H. Huxley: One of the kindest and truest men that it was ever my good fortune to know.

Life and Letters, vol. 2, 182

Rev. J. B. Innes: We had been speaking of the apparent contradiction of some supposed discoveries with the Book of Genesis; he said, "you are (it would have been more correct to say you ought to be) a theologian, I am a naturalist, the lines are separate. I endeavour to discover facts without considering what is said in the Book of Genesis. I do not attack

Moses, and I think Moses can take care of himself." To the same effect he wrote more recently, "I cannot remember that I ever published a word directly against religion or the clergy. . . ."

Life and Letters, vol. 2, 288–89

Leonard Jenyns: He occasionally visited me at my Vicarage, at Swaffham Bulbeck, and we made Entomological excursions together, sometimes in the Fens—that rich district yielding so many rare species of insects and plants—at other times in the woods and plantations of Bottisham Hall. He mostly used a sweeping net, with which he made a number of successful captures I had never made myself, though a constant resident in the neighbourhood.

Jenyns 1887, 44

Henry Lettington [Darwin's gardener]: I often wish he had something to do. He moons about in the garden, and I have seen him stand doing nothing before a flower for ten minutes at a time. If he only had something to do I really believe he would be better.

John Lubbock, Darwin-Wallace celebration 1908, Linnean Society of London, 57

John Lewis [carpenter in Down village]: But he was always a rare man for snuff, black snuff, that Lundy Foot. He kept it on the hall table, in a big tin that held near two quarts, and he'd be running in and out of the study twenty times a day for a go.

A visit to Darwin's village,
Evening News, 12 February 1909, 4

Harriet Martineau: The simple, childlike, painstaking, effective Charles Darwin, who established himself presently at the head of living English naturalists.

Chapman 1877, vol. 1, 268

Lady Dorothy Nevill: I am sending curious plants to experimentalize upon to Mr. Darwin. I am so pleased to help in any way the labours of such a man—it is quite an excitement for me in my quiet life, my intercourse with him— he promises to pay me a visit when in London. I am sure he will find I am the missing link between man and apes.

Nevill 1919, 56

Marianne North: I was asked by Mrs. Litchfield [Henrietta Darwin] to come and meet her father, Charles Darwin, who wanted to see me, but could not climb my stairs. He was, in my eyes, the greatest man living, the most truthful, as well as the most unselfish and modest, always trying to give others rather than himself the credit of his own great thoughts and work.... I was much flattered at his wishing to see me, and when he said he thought I ought not to attempt any representation of the vegetation of the world until I had seen and painted the Australian, which was so unlike that of any other country, I determined to take it as a royal command and to go at once.

Symonds 1894 vol. 2, 87

Charles Eliot Norton: We have seen Darwin several times during the last ten days. He is a delightful person from his simplicity, sweetness and strength.... His face is massive, with little beauty of feature but much of expression. He has a lively humour, and a cheerful, friendly manner.... His talk is not often memorable on account of brilliant or impressive sayings, but it is always the expression of the qualities of mind and heart which combine in such rare excellence in his genius.

Norton 1913, vol. 1, 309, 477

Joseph Parslow [Darwin's butler]: He was a very social, nice sort of gentleman, very joking and jolly indeed; a good husband and a good father and a most excellent master. Even his footmen used to stay with him as long as five years. They would rather stay with him than take a higher salary somewhere else. The cook came there while young and stayed till his death—nearly thirty years.

> Jordan 1922, vol. 1, 273

Alfred Lord Tennyson Aug. 17th. [1868] Farringford. Mr. Darwin called, and seemed to be very kindly, unworldly, and agreeable. A. said to him, "Your theory of Evolution does not make against Christianity": and Darwin answered, "No, certainly not."

> Tennyson 1898, vol. 3, 74

Kliment Timiriazev: A few minutes later, and quite unexpectedly, Darwin entered the room. . . . I was confronted with an impressive old man with a large grey beard, deep-sunken eyes, whose calm and gentle look made you forget about the scientist and think about the man. I couldn't help comparing him to an ancient sage or an Old Testament patriarch, a comparison which has often been quoted since.

> Timiriazev 2006, 51

Mark Twain: I do regard it as a very great compliment and a very high honor that that great mind, laboring for the whole human race, should rest itself on my books. I am proud that he should read himself to sleep with them.

Twain 1910, 33

John Tyndall: I allude to Mr Charles Darwin, the Abraham of scientific men—a searcher as obedient to the command of truth as was the patriarch to the command of God.

Tyndall 1871, 368

Alfred Russel Wallace: As to Darwin, I know exactly our relative positions, & my great inferiority to him. I compare myself to a Guerilla chief, very well for a skirmish or for a flank movement, & even able to sketch out the plan of a campaign, but reckless of communications & careless about Commissariat;—while Darwin is the great General, who can manoeuvre the largest army, & by attending to his lines of communication with an impregnable base of operations, & forgetting no detail of discipline, arms or supplies, leads his forces to victory. I feel truly thankful that Darwin had been studying the subject so many years before me,

& that I was not left to attempt & to fail, in the great work he has so admirably performed.

A. R. Wallace to Charles Kingsley,
7 May 1869, Wallace Letters Online

Victoria, Princess Royal of Russia: She was very much *au fait* at the "Origin". . . . She said after twice reading you she could not see her way as to the origin of four things; namely the world, species, man, or the black and white races. Did one of the latter come from the other, or both from some common stock? And she asked me what I was doing, and I explained that in recasting the "Principles" I had to give up the independent creation of each species. She said she fully understood my difficulty, for after your book "the old opinions had received a shake from which they never would recover."

Charles Lyell to Darwin,
16 January 1865, DCP 4746

Recollections by His Family

The way he brought us up is shown by a little story about my brother Leonard, which my father was fond of telling. He came into the drawing-room and found Leonard dancing about on the sofa, which was forbidden, for the sake of the springs, and said, "Oh, Lenny, Lenny, that's against all rules," and received for answer, "Then I think you'd better go out of the room."

Life and Letters, vol. 1, 134

It is a proof of the terms on which we were, and also of how much he was valued as a play-fellow, that one of his sons when about four years old tried to bribe him with sixpence to come and play in working hours. We all knew the sacredness of working time, but that any one should resist sixpence seemed an impossibility. . . . Another mark of his unbounded patience was the way in which we were suffered to make raids into the study when we had an absolute need of sticking-plaster, string, pins, scissors, stamps, foot-rule, or hammer. These and other such necessaries were always to be found in the study, and it was the only place where this was a certainty. We used to feel it wrong to go in during worktime; still, when the necessity was great, we did so. I remember his patient look when he said once, "Don't you think you could not come in again, I have been interrupted very often."

Life and Letters, vol. 1, 136

Our elder cousin, Julia Wedgwood, said that the only place in my father's and mother's house where you might be sure of not meeting a child, was the nursery, and as a matter of fact we did live with our parents far more than do most children. Many a time, even during his working hours, was a sick child tucked up on my father's sofa, to be quiet and safe and soothed by his presence.

Emma Darwin, vol. 1, 468

I remember my father entering the drawing room at Down, apparently seeking for someone, when I, then a schoolboy, was sitting on the sofa with the *Origin of Species* in my hands. He looked over my shoulder and said: "I bet you half a crown that you do not get to the end of that book."

Keynes 1943, 35

As a young lad I went up to my father when strolling about the lawn, and he, after, as I believe a kindly word or two, turned away as if quite incapable of carrying on any conversation. Then there suddenly shot through my mind the conviction that he wished he were no longer alive. Must there not have been a strained and weary expression in his face to have produced in these circumstances such an effect on a boy's mind?

L. Darwin 1929, 121

My father much enjoyed wandering slowly in the garden with my mother or some of his children, or making one of a party, sitting out on a bench on the lawn; he generally sat, however, on the grass, and I remember him often lying under one of the big lime-trees, with his head on the green mound at its foot.

Life and Letters, vol. 1, 116

He could not help personifying natural things. This feeling came out in abuse as well as in praise—*e.g.* of some seedlings—"The little beggars are doing just what I don't want them to." He would speak in a half-provoked, half-admiring way of the ingenuity of a Mimosa leaf in screwing itself out of a basin of water in which he had tried to fix it. One might see the same spirit in his way of speaking of Sundew, earthworms, &c.

Life and Letters, vol. 1, 117

He had a boy-like love of sweets, unluckily for himself, since he was constantly forbidden to take them. He was not particularly successful in keeping the "vows," as he called them, which he made against eating sweets, and never considered them binding unless he made them aloud.

Life and Letters, vol. 1, 118

After his lunch, he read the newspaper, lying on the sofa in the drawing-room. I think the paper was the only non-scientific matter which he read to himself. Everything else, novels, travels, history, was read aloud to him. . . . After he had read his paper, came his time for writing letters. These, as well as the MS. of his books, were written by him as he sat in a huge horse-hair chair by the fire, his paper supported on a board resting on the arms of the chair.

Life and Letters, vol. 1, 118–19

In money and business matters he was remarkably careful and exact. He kept accounts with great care, classifying them, and balancing at the end of the year like a merchant. I remember the quick way in which he would reach out for his account-book to enter each cheque paid, as though he were in a hurry to get it entered before he had forgotten it.

Life and Letters, vol. 1, 120

He had a pet economy in paper, but it was rather a hobby than a real economy. All the blank sheets of letters received were kept in a portfolio to be used in making notes; it was his respect for paper that made him write so much on the backs of his old MS., and in this way, unfortunately, he destroyed large parts of the original MS. of his books.

Life and Letters, vol. 1, 121

He took snuff for many years of his life, having learnt the habit at Edinburgh as a student. . . . He generally took snuff from a jar on the hall table, because having to go this distance for a pinch was a slight check; the clink of the lid of the snuff jar was a very familiar sound. Sometimes when he was in the drawing-room, it would occur to him that the study fire must be burning low, and when some of us offered to see after it, it would turn out that he also wished to get a pinch of snuff.

Life and Letters, vol. 1, 122

He would often lie on the sofa and listen to my mother playing the piano. He had not a good ear, yet in spite of this he had a true love of fine music. He used to lament that his enjoyment of music had become dulled with age, yet within my recollection his love of a good tune was strong. I never heard him hum more than one tune, the Welsh song "Ar hyd y nos," which he went through correctly; he used also, I believe, to hum a little Otaheitan song.

Life and Letters, vol. 1, 123

He was extremely fond of novels, and I remember well the way in which he would anticipate the pleasure of having a novel read to him, as he lay down, or lighted his cigarette. He took a vivid interest both in plot and characters, and would on no account know beforehand, how a story finished; he considered looking at the end of a novel as a feminine vice. He could not enjoy any story with a tragical end, for this reason he did not keenly appreciate George Eliot, though he often spoke warmly in praise of *Silas Marner*. Walter Scott,

Miss Austen, and Mrs. Gaskell, were read and re-read till they could be read no more.

Life and Letters, vol. 1, 124–25

Much of his scientific reading was in German, and this was a great labour to him. . . . When he began German long ago, he boasted of the fact (as he used to tell) to Sir J. Hooker, who replied, "Ah, my dear fellow, that's nothing; I've begun it many times."

Life and Letters, vol. 1, 126

In the non-biological sciences he felt keen sympathy with work of which he could not really judge. For instance, he used to read nearly the whole of *Nature,* though so much of it deals with mathematics and physics. I have often heard him say that he got a kind of satisfaction in reading articles which (according to himself) he could not understand.

Life and Letters, vol. 1, 127

Any public appearance, even of the most modest kind, was an effort to him.

Life and Letters, vol. 1, 128

His love of scenery remained fresh and strong. Every walk at Coniston [in the English Lake District] was a fresh delight, and he was never tired of praising the beauty of the broken hilly country at the head of the lake. One of the happy memories of this time [1879] is that of a delightful visit to Grasmere: "The perfect day," my sister [Henrietta Litchfield] writes, "and my father's vivid enjoyment and flow of spirits, form a picture in my mind that I like to think of. He could hardly sit still in the carriage for turning round and getting up to admire the view from each fresh point, and even in returning he was full of the beauty of Rydal Water, though he would not allow that Grasmere at all equalled his beloved Coniston."

Life and Letters, vol. 1, 129

He was always rejoiced to get home after his holidays; he used greatly to enjoy the welcome he got from his dog Polly, who would get wild with excitement, panting, squeaking, rushing round the room, and jumping on and off the chairs; and he used to stoop down, pressing her face to his, letting her lick him, and speaking to her with a peculiarly tender, caressing voice.

Life and Letters, vol. 1, 130

He was never quite comfortable except when utterly absorbed in his writing. He evidently dreaded idleness as robbing him of his one anodyne, work.

L. Darwin 1929, 120

Another quality which was shown in his experimental work, was his power of sticking to a subject; he used almost to apologise for his patience, saying that he could not bear to be beaten, as if this were rather a sign of weakness on his part. He often quoted the saying, "It's dogged as does it."

Life and Letters, vol. 1, 149

His courteous and conciliatory tone towards his reader is remarkable, and it must be partly this quality which revealed his personal sweetness of character to so many who had never seen him. . . . The tone of such a book as the "Origin" is charming, and almost pathetic; it is the tone of a man who, convinced of the truth of his own views, hardly expects to convince others.

Life and Letters, vol. 1, 155–56

The love of experiment was very strong in him, and I can remember the way he would say, "I shan't be easy till I have tried it," as if an outside force were driving him.

Life and Letters, vol. 1, 150

His love and goodness towards his little grandson Bernard were great; and he often spoke of the pleasure it was to him to see "his little face opposite to him" at luncheon. He and Bernard used to compare their tastes; *e.g.*, in liking brown sugar better than white, &c.; the result being, "We always agree, don't we?"
Life and Letters, vol. 1, 135

I have a vivid recollection of the pleasure of turning out my bottle of dead beetles for my father to name, and the excitement, in which he fully shared, when any of them proved to be uncommon ones.
Life and Letters, vol. 2, 140

He walked with a swinging action, using a stick heavily shod with iron, which he struck loudly against the ground, producing as he went round the "Sand-walk" at Down, a rhythmical click which is with all of us a very distinct remembrance.
Life and Letters, vol. 1, 109

Two peculiarities of his indoor dress were that he almost always wore a shawl over his shoulders, and that he had great loose cloth boots lined with fur which he could slip on over his indoor shoes. Like most delicate people he suffered from heat as well as from chilliness; it was as if he could not hit the balance between too hot and too cold.

Life and Letters, vol. 1, 112

In earlier times he took a certain number of turns [around the "Sand-walk"] every day, and used to count them by means of a heap of flints, one of which he kicked out on the path each time he passed. Of late years I think he did not keep to any fixed number of turns, but took as many as he felt strength for.

Life and Letters, vol. 1, 115

When I was at work on *Life and Letters* I had not seen it [Darwin's 1842 manuscript essay on natural selection]. It only came to light after my mother's death in 1896 when the house at Down was vacated. The MS. was hidden in a cupboard under the stairs which was not used for papers of any value, but rather as an overflow for matter which he did not wish to destroy.

F. Darwin 1909, xvii

My father [George Darwin] explained to me once, that my grandfather [Charles Darwin] was rather different from his children, because he was only half a Wedgwood, while they had a double dose of Wedgwood blood in them, owing to the two Darwin-Wedgwood marriages in two successive generations. "You've none of you ever seen a Darwin who wasn't mostly Wedgwood," he said, rather sadly, as of a dying strain.

Raverat 1952, 154

During the night of April 18th, about a quarter to twelve, he had a severe attack and passed into a faint, from which he was brought back to consciousness with great difficulty. He seemed to recognise the approach of death, and said, "I am not the least afraid to die." All the next morning he suffered from terrible nausea and faintness, and hardly rallied before the end came. He died at about four o'clock on Wednesday, April 19th, 1882.

Life and Letters, vol. 3, 358

Tributes

We hope you will not think we are taking a liberty if we venture to suggest that it would be acceptable to a very large number of our countrymen of all classes and opinions that our illustrious countryman, Mr. Darwin, should be buried in Westminster Abbey.

> John Lubbock, Memorial, to
> G. G. Bradley, Dean of Westminster,
> 21 April 1882, *Life and Letters*, vol. 1, 360

In 1859 was published what may be regarded as the most momentous of all his works, "The Origin of Species by means of Natural Selection." No one who had not reached manhood at the time can have any idea of the consternation caused by the publication of this work. We need not repeat the anathemas that were hurled at the head of the simple-minded observer, and the prophecies of ruin to religion and morality if Mr. Darwin's doctrines were accepted. No one, we are sure, would be more surprised than the author himself at the results which followed. But all this has long passed. . . . It has been said, perhaps prematurely, that one must seek back to Newton or even Copernicus, to find a man whose influence on human thought and methods of looking at the universe has been as radical as that of the naturalist who has just died.

Obituary, *The Times*, 21 April 1882, 5

Darwin has been read much, but talked about more. Since the publication of his work "On the Origin of Species" in 1859, and particularly within the 11 years which have elapsed since his "Descent of Man" was given to the world, he has been the most widely known of living thinkers . . . school children intuitively understood that if man is descended from the ape, he cannot be descended from Adam. All that part of the world which had never thought of such things before was aroused by the shock of the new idea.

Obituary, *New York Times*, 21 April 1882

He passed that life in elaborating one central idea, and he remained in the world long enough to see the whole course of modern science altered by his speculations.

Obituary, *Morning Post*, 21 April 1882

Very few, even among those who have taken the keenest interest in the progress of the revolution in natural knowledge set afoot by the publication of the *Origin of Species*; and who have watched, not without astonishment, the rapid and complete change which has been effected both inside and outside the boundaries of the scientific world in the attitude of men's minds towards the doctrines which are expounded in that great work, can have been prepared for the extraordinary manifestation of affectionate regard for the man, and of profound reverence for the philosopher, which followed the announcement, on Thursday last, of the death of Mr. Darwin.

T. H. Huxley, Obituary, *Nature*, 27 April 1882

We could not think, we could not bear to think, that that untiring and fertile brain, that simple, kindly heart, could cease to work in our midst for many a year yet to come. We looked forward to many another of the familiar green-bound volumes, rich with teeming facts and marvellous applications of minute discovery.

Grant Allen, Obituary, *The Academy* 21 (29 April 1882): 306

What human life could be more full
Of high achievement, as of years?
Who else has found so much to cull
Of ripened fruit that labour bears?
If life of man must have an end
Who would not gladly end like this?

George Romanes, quoted in
Pleins 2014, 329–30

Why did so many of the greatest intellects fail, while Darwin and myself hit upon the solution of this problem. . . . As I have found what seems to me a good and precise answer to this question, and one which is of some psychological interest, I will, with your permission, briefly state what it is. On a careful consideration, we find a curious series of correspondences, both in mind and in environment, which led Darwin and myself, alone among our contemporaries, to reach identically the same theory. First (and most important, as I believe), in early life both Darwin and myself became ardent beetle-hunters. Now there is certainly no group of organisms that so impresses the collector by the almost infinite number of its specific forms, the endless modifications of structure, shape, colour, and surface-markings that distinguish them from each other, and their innumerable adaptations to diverse environments.

> A. R. Wallace, Darwin-Wallace celebration 1908, Linnean Society of London, 7–8

You can be a thorough-going Neo-Darwinian without imagination, metaphysics, poetry, conscience, or decency. For "Natural Selection" has no moral significance: it deals with that part of evolution which has no purpose, no intelligence, and might more appropriately be called accidental selection, or better still, Unnatural Selection, since nothing is more unnatural than an accident. If it could be proved that the whole universe had been produced by such Selection, only fools and rascals could bear to live.

> George Bernard Shaw, *Back to Methuselah*, 1921, lxi–lxii

The theory of evolution is without any doubt the most important generalization yet made in the field of biology, worthy to rank with the great generalizations of the physical sciences, such as the conservation and degradation of energy, the modern theory of the atom, or Newton's theory of gravitation.

> Julian Huxley 1939, 1

Varia

He was born at Shrewsbury on February 12th, 1809. W. E. Gladstone, Alfred Tennyson, and Abraham Lincoln were born in the same year.

> Lankester 1896–97, vol. 2, 4835

Just as Darwin discovered the law of evolution in organic nature, so Marx discovered the law of evolution in human history.

> F. Engels, speech at the grave of
> Karl Marx, 17 March 1883,
> Marxist Internet Archive

She [Henrietta Darwin Litchfield] came to see me and asked what was the matter with me, I said "latent gout."

"Oh! That's what we have, does it come from drink in your parents?" It occurred to me that the Darwinian mind must be greater in science than in society.

> Alice James, *Diary*, 228

> The popular triumph of Darwinism must be the death-blow to theology.... Evolution and special creation are antagonistic ideas.
>
> Foote 1889, 4–5

> Humanity has in the course of time had to endure from the hands of science two great outrages upon its naive self-love. The first was when it realized that our earth was not the center of the universe, but only a tiny speck in a world-system of a magnitude hardly conceivable; this is associated in our minds with the name of Copernicus, although Alexandrian doctrines taught something very similar. The second was when biological research robbed man of his peculiar privilege of having been specially created, and relegated him to a descent from the animal world, implying an ineradicable animal nature in him: this transvaluation has been accomplished in our own time upon the instigation of Charles Darwin, Wallace, and their predecessors, and not without the most violent opposition from their contemporaries.
>
> Freud 1920, 246–47

> The first objection to Darwinism is that it is only a guess and was never anything more. It is called a "hypothesis," but the word "hy-

pothesis," though euphonious, dignified and high-sounding, is merely a scientific synonym for the old-fashioned word "guess." If Darwin had advanced his views as a *guess* they would not have survived for a year, but they have floated for half a century, buoyed up by the inflated word "hypothesis." When it is understood that "hypothesis" means "guess," people will inspect it more carefully before accepting it.

William Jennings Bryan, *New York Times*, 26 February 1922

For a lawyer, I was a fairly grounded scientist. I had been reared by my father on books of science. Huxley's books had been household guests with us for years, and we had all of Darwin's as fast as they were published.

Darrow 1932, 250

My dear, descended from the apes! Let us hope it is not true, but if it is, let us pray it does not become widely known.

Anecdote cited by Montagu 1942, 27

I first became aware of Charles Darwin and evolution while still a schoolboy growing up in Chicago . . . It is extraordinary the extent to which Darwin's insights not only changed his contemporaries' view of the world but also continue to be a source of great intellectual stimulation for scientists and nonscientists alike.

James D Watson, *Los Angeles Times*, 18 September 2005

It is not the strongest of the species that survives, nor the most intelligent. It is the one that is most adaptable to change.

Misattributed to Darwin

SOURCES

Short Titles

Autobiography: see Nora Barlow (ed.), 1958.
Beagle Diary: see R. D. Keynes (ed.), 1988.
Correspondence: see F. H. Burkhardt et al. (eds.), 1983–2016.
DCP: see Darwin Correspondence Project.
Darwin's Journal: see De Beer (ed.), 1959.
Descent: see Charles Darwin, 1871.
Emma Darwin: see Henrietta Litchfield (ed.), 1904.
Essay 1844: see Francis Darwin (ed.), 1909.
Expression: see Charles Darwin, 1872.
Journal of Researches 1839: see Charles Darwin, 1839.
Journal of Researches 1845: see Charles Darwin, 1845.
Life and Letters: see Francis Darwin (ed.), 1887.
More Letters: see Francis Darwin and A.C. Seward (eds.), 1903.
Notebook B, C, D, E, M, N: see Paul H. Barrett, et al., (eds.), 1987.
Orchids: see Charles Darwin, 1862.
Origin 1859: see Charles Darwin, 1859.
Origin 1861: see Charles Darwin, 1861.
Ornithological Notes: see Nora Barlow (ed.), 1963.
Variation: See Charles Darwin, 1868.

Agassiz, Elizabeth Cary (ed.). 1890. *Louis Agassiz: His life and correspondence.* Boston and New York: Houghton, Mifflin & Co.

Agassiz, Louis. 1860. [Review of] *On the Origin of Species. American Journal of Science and Arts* ser. 2, 30: 142–54.

Allingham, William. 1907. *William Allingham: A diary.* London: Macmillan and Co.

Aveling, E. B. 1883. *The religious views of Charles Darwin.* London: Freethought Publishing Company.

Barlow, Nora (ed.). 1958. *The autobiography of Charles Darwin 1809–1882. With the original omissions restored. Edited and with appendix and notes by his grand-daughter Nora Barlow.* London: Collins.

Barlow, Nora. 1963. Darwin's ornithological notes. *Bulletin of the British Museum (Natural History), Historical Series* 2 (7): 201–78.

Barrett, Paul H., et al. (eds.). 1987. *Charles Darwin's notebooks, 1836–1844: Geology, transmutation of species, metaphysical enquiries.* Cambridge: Cambridge University Press.

Burkhardt, F. H, et al. (eds). 1983–2016. *The correspondence of Charles Darwin.* Vols. 1–24 (1821–74). Cambridge, Cambridge University Press. See also Darwin Correspondence Project.

Chapman, M. W. 1877. *Harriet Martineau's autobiography.* 2 vols. Boston.

Cobbe, Frances Power. 1894. *Life of Frances Power Cobbe.* 2 vols. London: Richard Bentley & Son.

Conway, M. D. 1904. *Autobiography, memories and experiences.* 2 vols. London: Cassell and Company.

Darrow, Clarence. 1932. *The story of my life*. New York, Scribner's Sons.

Darwin Correspondence Project. https://www.darwinproject.ac.uk/.

Darwin Online. The Complete Work of Charles Darwin Online. http://darwin-online.org.uk/.

Darwin, Charles. 1839. *Journal of researches into the geology and natural history of the various countries visited by H.M.S. Beagle*. London: Colburn.

———. 1845. *Journal of researches into the natural history and geology of the countries visited during the voyage of H.M.S. Beagle round the world*. 2d ed. London: John Murray.

———. 1859. *On the origin of species by means of natural selection, or the preservation of favoured races in the struggle for life*. London: John Murray.

———. 1861. *On the origin of species by means of natural selection, or the preservation of favoured races in the struggle for life*. 3d ed. London: John Murray.

———. 1862. *On the various contrivances by which British and foreign orchids are fertilised by insects*. London: John Murray.

———. 1868. *The variation of animals and plants under domestication*. 2 vols. London: John Murray.

———. 1871. *The descent of man, and selection in relation to sex*. 2 vols. London: John Murray.

———. 1871. Pangenesis. *Nature: A Weekly Illustrated Journal of Science* 3 (27 April): 502–3.

———. 1872. *The expression of the emotions in man and animals*. London: John Murray.

Darwin, Charles. 1874. *The descent of man, and selection in relation to sex*. 2nd ed. London: John Murray.

———. 1877. A biographical sketch of an infant. *Mind: A Quarterly Review of Psychology and Philosophy* 2 (7): 285–94.

———. 1879. Preliminary notice. In Krause, E., *Erasmus Darwin*. Translated from the German by W. S. Dallas. London: John Murray.

Darwin, Charles, and A. R. Wallace. 1858. On the tendency of species to form varieties; and on the perpetuation of varieties and species by natural means of selection. *Journal of the Proceedings of the Linnean Society of London. Zoology* 3: 45–50.

Darwin, Francis (ed). 1887. *The life and letters of Charles Darwin, including an autobiographical chapter*. 3 vols. London: John Murray.

Darwin, Francis (ed.). 1909. *The foundations of The Origin of Species: Two essays written in 1842 and 1844*. Cambridge: Cambridge University Press.

Darwin, Francis. 1917. *Rustic Sounds and other studies in literature and natural history*. London: John Murray.

Darwin, Francis, and A. C. Seward (eds.). 1903. *More letters of Charles Darwin: A record of his work in a series of hitherto unpublished letters*. 2 vols. London: John Murray.

Darwin, Leonard. 1929. Memories of Down House. *The Nineteenth Century* 106:118–23.

De Beer, Gavin (ed.). 1959. *Darwin's Journal. Bulletin of the British Museum (Natural History) Historical Series* 2: 1–21.

FitzRoy, Robert, and Charles Darwin. 1836. A letter, containing remarks on the moral state of Tahiti, New Zealand, &c. *South African Christian Recorder* 2 (4): 221–38.

Foote, G. W. 1889. *Darwin on God*. London: Progressive Publishing Company.

Freud, Sigmund. 1920. *A general introduction to psychoanalysis*. Authorized translation, with a preface by G. Stanley Hall. New York: Boni and Liveright.

Galton, Francis. 1909. *Memories of my life*. New York: Dutton.

Gray, Asa. 1860. Darwin and his reviewers. *Atlantic Monthly* 6: 406–25.

Gotthelf, Allan. 1999. Darwin on Aristotle. *Journal of the History of Biology* 32: 3–30.

Gröben, Christiane (ed.). 1982. *Charles Darwin and Anton Dohrn, Correspondence*. Naples: Macchiaroli.

Gunther, A. E. 1975. The Darwin letters at Shrewsbury School. *Notes and Records of the Royal Society* 30: 25–43.

Haeckel, Ernst. 1882. On Darwin. *The Times*, 28 September, 6.

Healey, Edna. 2001. *Emma Darwin: The inspirational wife of a genius*. London: Headline.

Hodge, Charles. 1874. *What is Darwinism?* New York: Scribner.

Huxley, Julian. 1939. *The living thoughts of Darwin*. London: Cassell.

Huxley, Thomas Henry. 1860. The Origin of Species. *Westminster Review* 17 (n.s.): 541–70.

James, Alice. 1964. *The Diary of Alice James.* Edited by Leon Edel. London: Penguin Books.

Jensen, J. Vernon. 1988. Return to the Wilberforce-Huxley debate. *British Journal for the History of Science* 21: 161–79.

Jenyns, L. [L. Blomefield]. 1887. *Chapters in my life: With appendix containing special notices of particular incidents and persons.* Bath: privately printed.

Jordan, David Starr. 1922. *The days of a man: Being memories of a naturalist, teacher and minor prophet of democracy.* 2 vols. London: George Harrap.

Keynes, Margaret. 1943. Leonard Darwin. *Economic Journal* 53: 439–48.

Keynes, R. D. (ed.). 1988. *Charles Darwin's Beagle diary.* Cambridge: Cambridge University Press.

Lankester, E. Ray. 1896–97. Charles Robert Darwin. In *Library of the world's best literature ancient and modern*, ed. C. D. Warner. 30 vols. New York: Peale & Hill.

Litchfield, Henrietta (ed.). 1904. *Emma Darwin, wife of Charles Darwin. A century of family letters.* 2 vols. Cambridge: University Press.

Marx, Karl. 1975–2004. *Karl Marx, Frederick Engels: Collected works.* ed. Richard Dixon and others. 50 vols. New York: International Publishers.

Mill, John Stuart. 1862. *A system of logic, ratiocinative and inductive: Being a connected view of the principles of evidence and the methods of scientific investigation.* 5th ed. 2 vols. London.

Mivart, St. George J. 1871. [Review of] *The Descent of Man. Quarterly Review* 131 (July): 47–90.

Montagu, Ashley. 1942. *Man's most dangerous myth: The fallacy of race*. New York: Columbia University Press.

Morley, John. 1903. *The life of William Ewart Gladstone*. 2 vols. New York: Macmillan.

Nevill, Ralph. 1919. *The life and letters of Lady Dorothy Nevill*. London: Methuen & Co.

Norton, C. E. 1913. *Letters of Charles Eliot Norton*. 2 vols. Cambridge, MA: Houghton Mifflin.

Owen, Richard. 1860. Review of *On the Origin of Species*. *Edinburgh Review* 111: 487–532.

Pas, Peter W. van der. 1970. The correspondence of Hugo de Vries and Charles Darwin. *Janus* 57: 173–213.

Peart, Sandra J., and David M. Levy. 2008. Darwin's unpublished letter at the Bradlaugh-Besant trial: A question of divided expert judgement. *European Journal of Political Economy* 24: 343–53.

Raverat, Gwen. 1952. *Period piece: A Cambridge childhood*. London, Faber and Faber.

Symonds, J. C. (ed.). 1894. *Recollections of a happy life, being the autobiography of Marianne North*. 2 vols. New York: Macmillan.

Tennyson, Hallam (ed.) 1898. *The life and works of Alfred Lord Tennyson*. 10 vols. London: Macmillan and Co.

Timiriazev, K. A. 2006. A visit to Darwin, with notes by Leon Bell. *Archipelago* 9 (2006): 47–58.

Twain, Mark [Samuel Clemens]. 1910. *Mark Twain's speeches*. Ed. W. D. Howells. New York and London: Harper Brothers.

Tyndall, John. 1871. Fragments of Science for unscientific people: A series of detached essays, lectures, and reviews. New York: D. Appleton.

Wallace, A. R. 1869. Sir Charles Lyell on geological climates and the origin of species. *Quarterly Review* 126 (252): 359–94.

———. 1905. *My life: A record of events and opinions*. 2 vols. London: Chapman and Hall.

Wilberforce, Samuel. 1860. [Review of] *On the origin of species*. *Quarterly Review* 108: 225–64.

INDEX

aboriginals, Australian, 44, 200, 207–8
Abutilon darwinii, 169
The Academy, obituary of CD in, 300
adaptation, 65, 66, 70, 124–25, 305; beneficial, 166; and domesticated species, 122; and good of another species, 123; and Lamarck, 90; machinery metaphor for, 125; and morphology, 56, 67, 108–9, 111, 142; neither beneficial nor injurious, 142; number and diversity of inheritable deviations of, 138–39; as produced for good of possessor, 128; usefulness of, 114, 119
Admiralty, British, 33
Africa, 26
Agassiz, Elizabeth Cary, 263

Agassiz, Louis, 113, 158–59, 234, 263
agnosticism, 223
Allen, Grant, 300
Allingham, William, 275
Andaman islanders, 174
Andes Mountains, 26–27
Anglican Church, 277
Angracum sesquipedale (Darwin's orchid), 166–67
animals: affinities of, 54; checks on increase of, 57–58; classification of, 54; and death, 200; descent of, 144; domesticated, 53, 83, 122, 125, 138, 139, 140, 185, 186, 211 (*see also* mankind, power over selection); domesticated *vs.* wild, 40; extinct, 26; female, 185, 186–87; and human intellect, 193; humanity to, 247–48; and human moral sense, 190, 191;

animals (*cont.*)
  instincts of, 54; mankind's descent from, 304; migration of, 130–31; one no higher than another, 54; progenitors of, 144; and races of men, 181; and self-consciousness, 200; stronger as extirpating weaker, 208; and struggle for existence, 58; suffering of, 219–20; warring among, 66. *See also* birds; population; species
ants, 128–29, 198–99
ape(s), 74; brain of, 174; and human intellect, 193; mankind as descended from, 157, 158, 161, 177, 299, 305; and mirrors, 74, 204–5. *See also* chimpanzees; mandrills; monkeys; orangutans
architect, 136–37
Argus pheasant, 187
aristocracy, 233
Aristotle, 238, 239
armadillos, 69
Ascension Island, 22
astronomy, 134–35, 136
asylums, 210

atheism, 133, 164, 223, 265
atolls, 29
Austen, Jane, 292
Australia, 44, 207, 208
Australian settlers, 174
The Autobiography of Charles Darwin (C. Darwin): barnacles in, 84, 89; *Beagle* voyage in, 13, 14–15, 22–23; botany in, 167, 169; children in, 79; design in, 133; divergence with modification in, 70; education in, 3–10; embryology in, 164; emotion in, 202, 203; friends and contemporaries in, 263–64, 265, 266, 268, 269–73, 274; geology in, 24, 28–29; health issues in, 226; human origins in, 173; marriage in, 63–64; personal life in, 253–58, 259; precursors in, 90, 91, 95–96; religious belief in, 217–20, 222–23, 225; science in, 234–35, 239; slavery in, 30–31; species in, 53, 57, 58, 69; Wallace in, 102, 116; writing habits in, 241–42
Aveling, E. B.: CD's 13 Oc-

tober 1880 letter to, 223–24; *The religious views of Charles Darwin*, 224–25
Avestruz Petise, 33–34

Babbage, Charles, 264
baboons, 56, 258
Back to Methuselah (Shaw), 302
Baconian principles, 53
Bahia, Brazil, 30
Bajada [Baja de Entre Rios, River Parana, Argentina], 25
barnacles, 84–89, 152, 241
barn-owl, white, 65
Bartlett, Abraham D., CD's 5 January [1870] letter to, 202–3
Bateman, James, 166
Bates, H. W., 156, 241
Batrachians, 34
Bay of San Blas [south of Bahia Blanca, Argentina], 35
Beagle. See *HMS Beagle*
Beagle Channel, South America, 39
Beagle Diary (C. Darwin), 39, 41–42
bears, black, 111
Beaufort, Francis, 33
beauty, 128, 181, 186, 187
bees, 110, 190–91, 199–200
beetles, 6–7, 21, 76–77, 108, 295, 301
Bell, Thomas, 103
Benchuca bug, 19
Bible, 217, 218, 224, 279–80, 283, 284, 299
birds, 109, 115, 140, 252; and courtship, 185–87, 188; and Galápagos Archipelago, 46, 48, 49, 65; and instincts, 200; species of, 53–54; tameness of, 46; taxidermy of, 5. *See also* animals
bivalves, 133
Blyth, Edward, 97
botany, 165–69
Brachinus crepitans, 76
Brachiopod shells, genera of, 120, 121
Bradlaugh, Charles, CD's 6 June 1877 letter to, 192
Bradley, G. G., 298
brain, 55, 174, 182–83, 201
Brazil, 15, 32, 217–18
breeders. *See* animals, domesticated; mankind, power over selection; plants, domesticated
Bressa prize (Turin Society, Italy), 238
British Association for the Advancement of Science, 157, 161

British colonies, 208
British empire, xiii
Brodie, Sir Benjamin, 157
Brown, Robert, 57, 263
Bryan, William Jennings, 304–5
Büchner, Georg, 200
buds, 139, 141
Buffon, Georges-Louis Leclerc, Comte de, 95
Butler, Samuel (author), 263–64, 275
Butler, Samuel (headmaster, Shrewsbury School), 3
butterflies, 35
Button, Jemmy (Orundellico), 41, 42
Byron, Lord, 257

California, gold rush in, 208–9
Callithrix sciureus, 202–3
Cambrian formation, 163
Cambridge University, 5–6, 7–10
Cameron, Julia Margaret, 216, 256
Candolle, Alphonse de, 66, 118; CD's 6 July 1868 letter to, 174
Candolle, Augustin Pyramus de, 57
Cape of Good Hope, 207

Cape Verde islands, 24
Carlyle, Thomas, 264
Carus, J. V., letter to CD, 15 November 1866, 161
cats, 110, 134
celibacy, 192
cells, 139, 140
Chambers, Robert, *Vestiges of the Natural History of Creation*, 91, 93–94, 154
chance, 134, 139
Chapman, John, CD's 16 May [1865] letter to, 230–31
character/personality, 186–87, 255, 258; of CD (*see under* Darwin, Charles Robert); of E. A. Darwin, 265
chastity, 191–92
Chemical Catechism (Henry and Parkes), 4
chemistry, CD's love for, 3–4, 150, 265
Chile: barnacles from, 84; *Benchuca* of, 19; coral reefs in, 28–29; earthquake in, 18–19; volcanic eruption in, 27
chimpanzees, 204. *See also* apes
Chonos Archipelago, Chile, 35

Christianity, 43, 218–19, 224, 265, 276, 283
civilization, 42, 183, 195, 255; advance of, 209; and arts, 193–94, 211; and Australian aborigines, 44; and chastity, 192; inheritance of wealth and property in, 211; and New South Wales, Australia, 208; preservation of weak members of, 210–11; and progress, 213–14
Civil War, American, 232, 233
classics, 78–79
clergy, 280; and CD's possible occupation, 12, 16, 21, 217
climatal change, 130
clovers, 110
Clytus mysticus, 76
co-adaptations, 108
Cobbe, Frances Power, 248; CD's 23 March [1870] letter to, 190
cocoa-nut, 20
Coleridge, Samuel Taylor, 254, 257
collecting, 6, 11, 22, 33–38, 76–77, 80, 86, 88, 165, 301

Collier, John, CD's 16 February 1882 letter to, 258
colonialism, 32, 207
comets, 136
commerce, xiii
Concepción, Chile, 18–19
Concholepas, 84
Coniston, Lake District, 293
consciousness, 137
contraception, 192
controversies, CD's avoidance of, 256
Conway, Moncure Daniel, 276
Copernicus, Nicolai, 299, 304
Copley Medal of the Royal Society, 277
coral reefs, 28–29, 239
Covington, Syms: CD's 23 November 1850 letter to, 88, 208–9; CD's 30 March 1849 letter to, 86–87
creation, 113, 117; and Foote, 303; individual, 67; and Kingsley, 144; and Lyell, 285; and Powell, 94–95; special, 304; and Spencer, 94; as term, 145. *See also* God/Creator

Croll, James: CD's 19 September 1868 letter to, 163; CD's 31 January [1869] letter to, 163
Cuvier, Georges, 238–39

daisies *vs.* dandelions, 165
Dana, J. D.: CD's 8 October 1849 letter to, 87; CD's 29 September [1856] letter to, 80
Darrow, Clarence, 305
Darwin, Anne, 74–75
Darwin, Bernard Richard Meirion (grandson), 295
Darwin, Caroline Sarah (sister): CD's 30 March–12 April 1833 letter to, 39–40; CD's 13 November 1833 letter to, 18; CD's 13 October 1834 letter to, 18; CD's 10–13 March 1835 letter to, 18–19; CD's 29 April 1836 letter to, 240
Darwin, Charles Robert: appearance of, 16, 275, 276, 278–79, 282, 283, 295; birth of, 303; burial of, xiv, 298; caricature of, from *The Hornet*, 172; character of, 3, 12, 258, 277, 278, 281–83, 294; children of, xiv, 73–79, 286–88; daguerreotype of, 52; death of, 68, 297; education of, 3–10; last will and testament of, 259; life and habits of, 280, 281, 286–97; photograph by Cameron, 216, 256; photograph by Elliott and Fry, 262; photograph by Maull and Fox, 106; tastes of, 3, 7–8, 79, 254–55, 257, 258; tributes to, 298–302; watercolor sketch by Richmond, 2
—LETTERS: to E. B. Aveling, 13 October 1880, 224; to Abraham D. Bartlett 5 January [1870], 202–3; to Charles Bradlaugh, 6 June 1877, 192; to Alphonse de Candolle, 6 July 1868, 174; from J. V. Carus, 15 November 1866, 161; to John Chapman, 16 May [1865], 230–31; to Frances Power Cobbe, 23 March [1870], 190; to John Collier, 16 February 1882, 258; to Syms Covington, 30 March 1849, 86–87; to Syms Covington, 23 Novem-

ber 1850, 88, 208–9; to James Croll, 19 September 1868, 163; to James Croll, 31 January [1869], 163; to J. D. Dana, 8 October 1849, 87; to J. D. Dana, 29 September [1856], 80; to Caroline Sarah Darwin, 30 March–12 April 1833, 39–40; to Caroline Sarah Darwin, 13 November 1833, 18; to Caroline Sarah Darwin, 13 October 1834, 18; to Caroline Sarah Darwin, 10–13 March 1835, 18–19; to Caroline Sarah Darwin, 29 April 1836, 240; to Emily Catherine Darwin, 22 May–14 July 1833, 234; to Emily Catherine Darwin, 22 May [–14 July] 1833, 31–32; to Emily Catherine Darwin, 6 April 1834, 42; to Emily Catherine Darwin, 14 February 1836, 21, 207; from Emma Wedgwood Darwin, 21–22 November 1838, 61; from Emma Wedgwood Darwin, 23 January 1839, 61–62; from Emma Wedgwood Darwin, c. February 1839, 62; to Emma Wedgwood Darwin, 5 July 1844, 68; to Emma Wedgwood Darwin, 20–21 May 1848, 63; to Emma Wedgwood Darwin, [23 April 1851], 75; to Emma Wedgwood Darwin, [28 April 1858], 251–52; from Erasmus Alvey Darwin, 23 November [1859], 150–51; to Horace Darwin, [15 December 1871], 237; to Robert Waring Darwin, 31 August [1831], 11–12; to Robert Waring Darwin, 8 February–1 March [1832], 13–14; to Robert Waring Darwin, 10 February 1832, 14; to Susan Elizabeth Darwin, 14 July–7 August [1832], 16; to Susan Elizabeth Darwin, 4 August [1836], 21; to William Erasmus Darwin, 3 October [1851], 75; to Anton Dohrn, 4 January 1870, 236; to Anton Dohrn, 15 February 1880, 238; to *Entomolo-*

Darwin, Charles Robert — *LETTERS (cont.)*
  gist's *Weekly Intelligencer*, 25 June 1859, 76–77; to T. H. Farrer, 29 October [1868], 141; to Henry Fawcett, 18 September [1861], 235, 240; to Robert FitzRoy, [4 or 11 October 1831], 13; to Robert FitzRoy, [19 September 1831], 33; to Robert FitzRoy, 1 October 1846, 255; to John Fordyce, 7 May 1879, 223; to W. D. Fox, May [1832], 25; to W. D. Fox, [12–13] November 1832, 16; to W. D. Fox, [9–12 August] 1835, 21; to W. D. Fox, 15 February 1836, 266; to W. D. Fox, [7 June 1840], 73; to W. D. Fox, [24 April 1845], 91; to W. D. Fox, 17 July [1853], 78–79; to W. D. Fox, 3 October [1856], 80; to W. D. Fox, 8 February [1857], 71, 75–76; to W. D. Fox, 13 November [1858], 76, 228; to W. D. Fox, 24 [March 1859], 71, 255; from W. D. Fox, 28 November [1864], 277; to Francis Galton, 23 December [1869], 196; to Asa Gray, 20 July [1857], 71; to Asa Gray, 4 July 1858, 99; to Asa Gray, 11 November [1859], 72; to Asa Gray, [8 or 9 February 1860], 127; to Asa Gray, 3 April [1860], 127; to Asa Gray, 22 May [1860], 133–34; to Asa Gray, 3 July [1860], 135, 159; to Asa Gray, 22 July [1860], 267; to Asa Gray, 10 September [1860], 267; to Asa Gray, 5 June [1861], 232; to Asa Gray, 10–20 June [1862], 77, 162, 232–33; to Asa Gray, 23[–24] July [1862], 167; to Asa Gray, 28 May [1864], 255; to Asa Gray, 28 January 1876, 257; to Albert Gunther, 12 April [1874?], 37–38; from J. S. Henslow, 24 August 1831, 11; to J. S. Henslow, [5 September 1831], 12; to J. S. Henslow, [26 October–] 24 November 1832, 34; to J. S. Henslow, 18 April 1835, 249; to J. S.

INDEX 323

Henslow, 6 May 1849, 227; to J. M. Herbert, 2 June 1833, 32; to Frithiof Holmgren, 18 April 1881, 247–48; to J. D. Hooker, [11 January 1844], 90; to J. D. Hooker, 11 January 1844, 65; to J. D. Hooker, [7 January 1845], 91; to J. D. Hooker, [16 April 1845], 165; to J. D. Hooker, 11–12 July 1845, 68; to J. D. Hooker, 10 September 1845, 68–69; to J. D. Hooker, [3 September 1846], 165; to J. D. Hooker, [2 October 1846], 84–85; to J. D. Hooker, [6 November 1846], 85; to J. D. Hooker, 10 May 1848, 85, 86; to J. D. Hooker, 13 June [1850], 88; to J. D. Hooker, 5 June [1855], 165; to J. D. Hooker, 13 July [1856], 70; to J. D. Hooker, 15 January [1858], 227–28; to J. D. Hooker, [29 June 1858], 76, 99; to J. D. Hooker, 13 [July 1858], 100–101; to J. D. Hooker, 1 September [1859], 228; to J. D. Hooker, 30 May [1860], 158–59; to J. D. Hooker, 4 December [1860], 168; to J. D. Hooker, 15 January [1861], 269; to J. D. Hooker, 23 [April 1861], 229; to J. D. Hooker, 19 June [1861], 165; to J. D. Hooker, 30 January [1862], 166–67; to J. D. Hooker, 9 February [1862], 235; to J. D. Hooker, 9 [April 1862], 229–30; to J. D. Hooker, 24 December [1862], 233; to J. D. Hooker, 15 February [1863], 168; to J. D. Hooker, 26 [March 1863], 236; to J. D. Hooker, [29 March 1863], 145; to J. D. Hooker, [27 January 1864], 168; to J. D. Hooker, 26[–27] March [1864], 230; from J. D. Hooker to CD, [11 June 1864], 279; to J. D. Hooker, 9 February [1865], 252; to J. D. Hooker, [29 July 1865], 270; to J. D. Hooker, 25 December [1868], 269; to J. D. Hooker, 16 January

Darwin, Charles Robert
—*LETTERS* (*cont.*)
[1869], 163; to J. D. Hooker, 1 February [1871], 145–46; to J. D. Hooker, 23 July [1871], 169; to J. D. Hooker, 18 January [1874], 237–38; to Leonard Horner 29 August [1844], 29; from A. von Humboldt, 18 September 1839, 279; to T. H. Huxley, 26 September [1857], 71; to T. H. Huxley, 2 June [1859], 71–72; to T. H. Huxley, 27 November [1859], 80–81; to T. H. Huxley, 28 December [1859], 152; to T. H. Huxley, 3 July [1860], 159, 270; to T. H. Huxley, [5 July 1860], 158; to T. H. Huxley, 22 November [1860], 160; to T. H. Huxley, 22 February [1861], 228–29; to T. H. Huxley, 30 January [1868], 202; to Henrietta Darwin Litchfield, [8 February 1870], 242; to Henrietta Darwin Litchfield, [March] 1870, 243; to Henrietta Darwin Litchfield, 20 March 1871, 179–80; to Henrietta Darwin Litchfield, 4 September 1871, 63; to Henrietta Darwin Litchfield, 4 September [1871], 78; to Henrietta Darwin Litchfield, 4 January 1875, 247; to John Lubbock, 5 September [1862], 77, 235; to Charles Lyell, 30 July 1837, 53–54; to Charles Lyell, [14] September [1838], 54; to Charles Lyell, [2 September 1849], 87; to Charles Lyell, 4 November [1855], 81; from Charles Lyell, 1–2 May 1856, 71, 80; to Charles Lyell, 18 [June 1858], 98; to Charles Lyell, [25 June 1858], 98–99; to Charles Lyell, 20 September [1859], 72; to Charles Lyell, [10 December 1859], 151–52; to Charles Lyell, 10 April [1860], 92, 155, 156, 272; to Charles Lyell, 4 May [1860], 174, 209; to Charles Lyell, 17 June [1860], 124, 134–35; to Charles Lyell, 20 [June

1860], 165; to Charles Lyell, 3 October [1860], 120, 124; to Charles Lyell, 14 November [1860], 168; to Charles Lyell, [1 August 1861], 136; to Charles Lyell, 12–13 March [1863], 95; from Charles Lyell, 15 March 1863, 161; from Charles Lyell, 16 January 1865, 285; to Victor A. E. G. Marshall, 7 [September] 1879, 272; to Frederick McDermott, 24 November 1880, 224; to N. A. von Mengden, 8 April 1879, 223; to J. F. T. Muller, [before 10 December 1866], 267; to John Murray, 14 June [1859], 240; to John Murray, [3 November 1859], 150; to William Ogle, 6 March [1868], 141; to William Ogle, 22 February 1882, 238–39; to Baden Powell, 18 January [1860], 91–92; to G. J. Romanes, 7 March 1881, 197; from Adam Sedgwick, 24 November 1859, 151; from A.R. Wallace, 2 July 1866, 125; to A. R. Wallace, 22 December 1857, 97, 173, 235; to A. R. Wallace, 18 May 1860, 157; to A. R. Wallace, 28 [May 1864], 181, 233; to A. R. Wallace, 15 June [1864], 182; to A. R. Wallace, 5 July [1866], 147–48; to A. R. Wallace, 27 March [1869], 142, 175; to A. R. Wallace, 14 April 1869, 175; to A. R. Wallace, 20 April [1870], 273; to Julia Wedgwood, 1 July [1861], 135–36; to C. T. Whitley, 8 May 1838, 59

—WORKS: *The Autobiography of Charles Darwin*, 3–10, 13, 14–15, 22–23, 24, 28–29, 30–31, 53, 57, 58, 63–64, 69, 70, 79, 84, 89, 90, 91, 95–96, 102, 116, 133, 164, 167, 169, 173, 202, 203, 217–20, 222–23, 225, 226, 234–35, 239, 241–42, 253–58, 259, 263–64, 265, 266, 268, 269–73, 274; *Beagle Diary*, 39, 41–42; *The Descent of Man* (1871), 142, 175, 176–78, 179, 182–84, 185–88, 190–92, 193–97,

Darwin, Charles Robert
—WORKS (cont.)
200, 203, 209–13, 220–22, 233, 236–37, 242, 243, 246, 258–59, 299; *The Descent of Man* (1874), 188–89, 200–201, 213–14; *The Expression of the Emotions in Man and Animals*, 74, 137, 204–6, 245–46; *Journal*, 61, 70, 88; *Journal of Researches* (1839), xii, 15–18, 19–21, 25–28, 31, 33–37, 44, 45–48, 54, 207–8, 218, 249–51; *Journal of Researches* (1845), 49; "A letter, containing remarks," 1836, 40, 43–44; *Notebook B*, 54, 55; *Notebook C*, 55; *Notebook D*, 56, 57–58; *Notebook E*, 181; *Notebook M*, 56, 256; *Notebook N*, 57; *On the Origin of Species* (1859), xi, xii, xvi, 82–83, 89, 95–96, 102, 107–16, 117–20, 121, 122–23, 127, 128–32, 138–39, 142, 144, 150–64, 173, 198–200, 222, 240, 241, 277–78, 287–88, 294, 298–99, 300; *On the Origin of Species* (1861), 92–95, 144–45; *On the Origin of Species* (1869), 142; *On The Tendency of Species to Form Varieties; and on the Perpetuation of Varieties and Species by Natural Means Of Selection*, 100; *On the Various Contrivances by which British and Foreign Orchids are Fertilised by Insects*, 124–25, 165–67; *Ornithological Notes*, 48; Preface to *Life of Erasmus Darwin*, 275; *The Variation of Animals and Plants under Domestication*, 83, 125–26, 136–37, 139–41, 142, 148–49

Darwin, Charles Waring (son), 75–76; death of, 76

Darwin, Emily Catherine (sister): CD's 22 May–14 July 1833 letter to, 234; CD's 22 May [–14 July] 1833 letter to, 31–32; CD's 6 April 1834 letter to, 42; CD's 14 February 1836 letter to, 21, 207

Darwin, Emma Wedgwood (wife), 235; and backgammon, 257; on CD and religion, 223, 225; CD's 5 July 1844 letter to, 68; CD's 20–21 May 1848 letter to, 63;

CD's [23 April 1851] letter to, 75; CD's [28 April 1858] letter to, 251–52; and CD's enjoyment of nature, 251–52; CD's love for, 61–64; and children, 287; and death of Anne, 75; and garden walks, 288; health of, 227; letter to CD, 21–22 November 1838, 61; letter to CD, 23 January 1839, 61–62; letter to CD, c. February 1839, 62; letter to Francis Darwin about CD's religious statement in *Autobiography*, 225; marriage to CD, xiv, 59–64; and music, 291; and publication of CD's species theory, 68; and visit of Mrs. Huxley, 228–29

Darwin, Erasmus (grandfather), 95, 256, 265, 278; *Zoonomia*, 90, 93

Darwin, Erasmus Alvey (brother), 3–4, 56, 92, 219, 237, 264, 265; 23 November [1859] letter to CD, 150–51

Darwin, Francis (son), 76–77, 89, 160, 240–41, 244–45, 296; letter from Emma Wedgwood Darwin to, 225; *The Life and Letters of Charles Darwin*, 258, 286–87, 288–93, 294–96, 297

Darwin, George Howard (son), 75, 227, 237, 238, 296–97

Darwin, Henrietta Emma (daughter). *See* Litchfield, Henrietta Emma Darwin (daughter)

Darwin, Horace (son), 76–77; CD's [15 December 1871] letter to, 237

Darwin, Leonard (son), 76–77, 227, 286, 288, 294

Darwin, Robert Waring (father), 22, 217, 219, 253–54; CD's 31 August [1831] letter to, 11–12; CD's 8 February–1 March [1832] letter to, 13–14; CD's 10 February 1832 letter to, 14; objections of to CD's *Beagle* journey, 11–12

Darwin, Susan Elizabeth (sister): CD's 14 July–7 August [1832] letter to, 16; CD's 4 August [1836] letter to, 21

Darwin, Susanna Wedgwood (mother), 253

Darwin, William Erasmus (son), 73–74; CD's 3 October [1851] letter to, 75; daguerreotype of, 52
Daubeny, C.G.B., 157
death, 210, 250, 252, 288
deduction, 58, 239, 273
Demosthenes, 236
Descartes, René, 273
descent, 68; and Lyell, 271; theory of, 92. *See also* natural selection
descent, of mankind, 56, 93, 175, 258–59, 304; from apes, 157, 158, 177, 299, 305; from lower form, 176–78
descent, with modification, xiii, 65–72, 129, 131
The Descent of Man (1871) (C. Darwin), 299; adaptation in, 142; descent in, 176–78, 258–59; dogs in, 246; emotions in, 203; instincts in, 200; intellect in, 193–96; morality in, 190–92; natural selection in, 175; race in, 182–84; reception of, 179; religion in, 220–21; science in, 236–37; sexual selection in, 185–88; slavery in, 233; society in, 209–13; style of, 242, 243
The Descent of Man (1874) (C. Darwin): instincts in, 200–201; sexual selection in, 188–89; society in, 213–14
design, xvi, 113, 133–37, 164. *See also* God/Creator; nature
De Vries, Hugo, 276
diamond beetle, 21
disease, 141, 181, 210
Disraeli, Benjamin, 161
dogs, 17, 59, 134, 183, 200, 203, 221, 244–48, 253, 258, 293
Dohrn, Anton, 276; CD's 4 January 1870 letter to, 236; CD's 15 February 1880 letter to, 238
D'Orbigny, Alcide, 34
doves, 46
Drosera, 167–68
Duncan, Andrew, 4–5
Dyster, Frederick Daniel, 158

earth, 252, 304; age of, 67, 163; destruction of, 222. *See also* geology
earthquakes, 27–28
earthworms, 289
Edinburgh Hospital, operation, 5
Edinburgh Review, 155

Edinburgh University, 6
education, 3–10, 23, 78–79, 214
Eliot, George, *Silas Marner*, 291
Elwin, Whitwell, 81; letter to John Murray, 21 September 1870, 176
embryology, 112–13, 164
emigration, 209, 210, 212
Emma Darwin (Litchfield), 61
emotions, 202–6, 221, 244
Engels, Friedrich, 162, 303
England, 31–32, 40, 207
Das entdeckte Geheimnis der Natur (Sprengel), 57
Entomologist's Weekly Intelligencer, CD's 25 June 1859 letter to, 76–77
An Essay on the Principle of Population (Malthus), 58
"Essays" (Spencer), 94
"Essays on the spirit of inductive philosophy, unity of worlds, and the philosophy of creation" (Powell), 94–95
Euclid, 3, 8
Europe, 123, 212
Europeans, 41, 207
Evidences of Christianity (Paley), 8–9
evolution, xi, xii, xvi, 223, 302, 303. *See also* descent, with modification
The Excursion (Wordsworth), 254
experimentation, xiii, 4, 235, 236, 294. *See also* observation
The Expression of the Emotions in Man and Animals (C. Darwin), 74, 137, 204–6, 245–46
eye, 127

facial expressions, 202, 204–5
facts, false, 237
Falconer, Hugh, 276–77
famine, 57–58
Farrer, T. H., CD's 29 October [1868] letter to, 141
Fawcett, Henry, CD's 18 September [1861] letter to, 235, 240
feathers, 127
fertilisation, 57, 110, 139, 165–66, 169
Ffinden, Rev. George, 277
finches, 46, 82
Fitzroy, Mary, 255
FitzRoy, Robert, xi, 11, 12, 13, 15, 16, 28, 37, 39, 42, 161, 266; CD's [19 September 1831] letter to, 33; CD's [4 or 11 Octo-

FitzRoy (*cont.*)
ber 1831] letter to, 13; CD's 1 October 1846 letter to, 255; "A letter, containing remarks," 1836, 40, 43–44; and slavery, 30–31

flowers, 15, 21, 57, 110

Foote, G. W., 304

Forbes, Edward, 91

Forbes, J. D., 160

Fordyce, John, CD's 7 May 1879 letter to, 223

forests, 217–18

Formica sanguinea, 198–99

fossils, 22, 25–26, 54, 69, 88, 150

Fox, William Darwin (cousin), 244, 266, 277; CD's May [1832] letter to, 25; CD's [12–13] November 1832 letter to, 16; CD's [9–12 August] 1835 letter to, 21; CD's 15 February 1836 letter to, 266; CD's [7 June 1840] letter to, 73; CD's [24 April 1845] letter to, 91; CD's 17 July [1853] letter to, 78–79; CD's 3 October [1856] letter to, 80; CD's 8 February [1857] letter to, 71, 75–76; CD's 13 November [1858] letter to, 76, 228; CD's 24 [March 1859] letter to, 71, 255; letter to CD, 28 November [1864], 277

foxes, 35, 48, 65

French government, 34

Freud, Sigmund, 304

Fuegia Basket (Yokcushla), 42

Fuegians, 39–40, 44, 178

Galápagos Archipelago, 37, 45–49, 54, 65, 69, 165

Galton, Francis, 143, 197, 213, 238, 277–78; CD's 23 December [1869] letter to, 196; *Hereditary Genius*, 196

Gaskell, Elizabeth, 292

gauchos, 16–17, 33

gemmules, 139–40, 245

gender, 185, 187–88, 191–92, 194–96, 212, 233

genealogical trees, 71

genera, 70, 108; protean or polymorphic, 120–21

genetics. *See* variation/variability

geography, 3, 120, 130–31, 150

Geological Society of London, 26

geological time, 120, 121, 122–23, 129, 131
geology, xii, 9, 14–15, 24–29, 53, 54, 67, 69, 129; and Chambers, 91; and Croll, 163; Jameson's lectures on, 4–5; and South American inhabitants, 107. *See also* earth
geometry, 3
German language, 292
glacier-lake theory, 234
Gladstone, William Ewart, 266, 303
Glen Roy, Scotland, 234
God/Creator, xi, 37, 55, 66, 72, 112, 114, 133–34, 135, 136, 137, 166, 218, 224; and Erasmus Darwin, 265; debates over, xiii; and design, 133, 164; and Kingsley, 144; limited belief in, 220; and original created forms, 144–45; and sense of sublimity, 222–23; and suffering, 219–20; and theory of evolution, 223. *See also* design; religion
Goethe, Johann Wolfgang von, 93, 278
Gould, John, 55
Grant, Robert, 90
Grasmere, 293
grasses, 15, 166
gravity, 55, 116, 125, 134, 145. *See also* scientific laws
Gray, Asa, 102, 156, 158, 164, 267; CD's 20 July [1857] letter to, 71; CD's 4 July 1858 letter to, 99; CD's 11 November [1859] letter to, 72; CD's [8 or 9 February 1860] letter to, 127; CD's 3 April [1860] letter to, 127; CD's 22 May [1860] letter to, 133–34; CD's 3 July [1860] letter to, 135, 159; CD's 22 July [1860] letter to, 267; CD's 10 September [1860] letter to, 267; CD's 5 June [1861] letter to, 232; CD's 10–20 June [1862] letter to, 77, 162, 232–33; CD's 23[–24] July [1862] letter to, 167; CD's 28 May [1864] letter to, 255; CD's 28 January 1876 letter to, 257
Gray, Jane Loring, 257
Gray, John Edward, 38
Gray, Thomas, 257
Great Britain, 210
Gully, William, 227

Gunther, Albert, CD's 12 April [1874?] letter to, 37–38

Haeckel, Ernst, 267, 278–79
Haiti/Hayti, 32
Handel, George Frideric, *Messiah*, 251
Haughton, Samuel, 102
hawks, 46
health, xiv, 18, 63, 88, 97, 226–31, 255, 256, 276, 295–96, 297. *See also* medicine
Hearne, Samuel, 111
Henry, William, *Chemical Catechism*, 4
Henslow, Rev. John Stevens, 9, 22, 24, 229, 268; 24 August 1831 letter to CD, 11; CD's [5 September 1831] letter to, 12; CD's [26 October–] 24 November 1832 letter to, 34; CD's 18 April 1835 letter to, 249; CD's 6 May 1849 letter to, 227
Herbert, J. M., CD's 2 June 1833 letter to, 32
Herbert, John Rogers (artist), 279
Hereditary Genius (Galton), 196

heredity, 63, 108, 137, 138–43, 194
hermaphrodites, 85, 86, 87, 89, 169
Herschel, Sir John, 95, 151–52; *A Preliminary Discourse on the Study of Natural Philosophy*, 10
Hieracium, 120
Hippocrates, 141
history, 3
hive-bees, 190–91, 199–200
HMS Beagle, xi, xiv, xvi, 11–23, 24, 40, 43, 69, 107, 217, 226, 244
Hodge, Charles, 164
Holmgren, Frithiof, CD's 18 April 1881 letter to, 247–48
home-sickness, 18, 21
Hooker, Joseph Dalton, 71, 100; and CD as forestalled, 100–101; and CD on *Abutilon darwinii*, 169; and CD on Chambers, 91; and CD on cruelty of nature, 70; and CD on death of son, 76; and CD on descent, 68; and CD on Drosera, 168; and CD on end of earth, 252; and CD on experimentation, 235, 236; and CD on hermaphrodites,

INDEX 333

85, 86; and CD on Jenkin, 163; and CD on Lamarck, 90; and CD on observation, 85; and CD on orchids, 165, 166–67; and CD on primordial soup, 145–46; and CD on priority, 99; and CD on species, 68–69; and CD on spiritualism, 237–38; and CD on variation, 88; CD's disagreements with, 96; CD's friendship with, 269; and CD's health, 227–28, 229–30; and CD's hothouse, 168; and CD's use of creation as term, 145; character of, 268; and German, 292; legacy for, 259; CD's [11 January 1844] letter to, 90; CD's 11 January 1844 letter to, 65; CD's [7 January 1845] letter to, 91; CD's [16 April 1845] letter to, 165; CD's 11–12 July 1845 letter to, 68; CD's 10 September 1845 letter to, 68–69; CD's [3 September 1846] letter to, 165; CD's [2 October 1846] letter to, 84–85; CD's [6 November 1846] letter to, 85; CD's 10 May 1848 letter to, 85, 86; CD's 13 June [1850] letter to, 88; CD's 5 June [1855] letter to, 165; CD's 13 July [1856] letter to, 70; CD's 15 January [1858] letter to, 227–28; CD's [29 June 1858] letter to, 76, 99; CD's 13 [July 1858] letter to, 100–101; CD's 23 January [1859] letter to, 101; CD's 1 September [1859] letter to, 228; CD's 30 May [1860] letter to, 158–59; CD's 4 December [1860] letter to, 168; CD's 15 January [1861] letter to, 269; CD's 23 [April 1861] letter to, 229; CD's 19 June [1861] letter to, 165; CD's 30 January [1862] letter to, 166–67; CD's 9 February [1862] letter to, 235; CD's 9 [April 1862] letter to, 229–30; CD's 24 December [1862] letter to, 233; CD's 15 February [1863] letter to, 168; CD's 26 [March 1863] letter to, 236; CD's [29 March 1863] letter to,

Hooker, Joseph Dalton (*cont.*)
145; CD's [27 January 1864] letter to, 168; CD's 26[–27] March [1864] letter to, 230; letter of [11 June 1864] to CD from, 279; CD's 9 February [1865] letter to, 252; CD's [29 July 1865] letter to, 270; CD's 25 December [1868] letter to, 269; CD's 16 January [1869] letter to, 163; CD's 1 February [1871] letter to, 145–46; CD's 23 July [1871] letter to, 169; CD's 18 January [1874] letter to, 237–38; and Wallace, 102; Wallace's 6 October 1858 letter to, 101

Hope, T. C., 4

Horace, 274

Horner, Leonard, CD's 29 August [1844] letter to, 29

horses, 25

humble-bees, 110

Humboldt, Alexander von, 269; letter to CD, 18 September 1839, 279; *Personal Narrative*, xii, 10

hunting, 6

Huxley, Henrietta Anne Heathorn, 202, 228–29

Huxley, Julian, 302

Huxley, Thomas Henry, 71, 153, 154, 157; character of, 269–70; on CD, 279; CD's legacy for, 259; CD's 26 September [1857] letter to, 71; CD's 2 June [1859] letter to, 71–72; CD's 27 November [1859] letter to, 80–81; CD's 28 December [1859] letter to, 152; CD's 3 July [1860] letter to, 159, 270; CD's [5 July 1860] letter to, 158; CD's 22 November [1860] letter to, 160; CD's 22 February [1861] letter to, 228–29; CD's 30 January [1868] letter to, 202; and Darrow, 305; letter to F. Dyster, 9 September 1860, 158; *Nature* obituary of CD by, 300

hypothesis, 142, 162, 239, 304–5. *See also* science

Ichneumonida, 134

iguana, land, 48

iguana, marine, 47

imagination, 221, 302

immortality, 221–22

indigenous peoples, 39–44, 176, 178, 193–94, 233, 255, 258–59; and belief in God, 220; elimination of weak, 210; and Fuegians as savages, 39–40; and gender, 187–88; mental requirements of, 174; and sexual selection, 181; skulls of, 183
induction, 151, 162
industrialization, xiii
innate qualities, 10
Innes, J. B., 279–80
insects, 15, 55, 57, 65, 109, 111, 115, 120, 167, 235, 280
instincts, 47, 54, 67, 73, 114, 128–29, 132, 148, 181, 191, 198–201, 207, 221, 246, 255
intellect, 55, 178, 183, 193–97, 211–12, 213–14, 221
Iquique, Chile, 19

James, Alice, *Diary*, 303
Jameson, Robert, 4–5
Jenkin, Fleeming, 163
Jenny (orangutan at London zoo), 56–57
Jenyns, Leonard, 280
Journal (C. Darwin), 61, 70, 88
Journal of Researches (C. Darwin, 1839), xii; *Beagle* voyage in, 15–18, 19–21; and Galápagos Archipelago, 45–48; geology in, 25–28; and indigenous peoples, 44; and natural history collecting, 33–37; nature in, 249–51; religion in, 218; and slavery, 31; society in, 207–8; and transmutation of species, 54
Journal of Researches (C. Darwin, 1845), 49

Kant, Immanuel, 190
King's College Chapel, Cambridge, 7–8
Kingsley, Rev. Charles, 144; A. R. Wallace's 7 May 1869 letter to, 284–85
Kirby, William, 37
Krause, Ernst, *Life of Erasmus Darwin*, 263–64

lacrymal glands, 204
Lake District, UK, 292–93. *See also* Wordsworth, William
Lamarck, Jean-Baptiste, 90, 92–93, 95, 154, 161
Lamb, Charles, 265
language, 177, 193

Lavater, Johann Kaspar, 13
laws. *See* scientific laws
Lawson, Mr. (resident on Galapagos), 45
Leibnitz, Gottfried Wilhelm, 273
"A letter, containing remarks," 1836 (FitzRoy and Darwin), 40, 43–44
Lettington, Henry (CD's gardener), 280
Lewis, John (carpenter in Downe village), 281
Licinus beetle, 76
life, origin of, 144–46; as hidden from man, 67; as mystery of mysteries, 49
The Life and Letters of Charles Darwin (F. Darwin, ed.), 258, 286–87, 288–93, 294–96, 297
Life of Erasmus Darwin (Krause), 263–64
Lincoln, Abraham, 303
Linnaeus, Carl, 238–39
Linnean Society of London, 99, 100, 102, 103, 229–30, 258, 280, 302
lion-ant, 37
Litchfield, Henrietta Emma Darwin (daughter), 78, 227, 281, 287, 293, 305; CD's [March] 1870 letter to, 243; CD's [8 February 1870] letter to, 242; CD's 20 March 1871 letter to, 179–80; CD's 4 September 1871 letter to, 63; CD's 4 September [1871] letter to, 78; CD's 4 January 1875 letter to, 247; *Emma Darwin*, 61
Litchfield, Richard, 63
lizards, 46–47, 48
Locke, John, 56
love, 221
Lubbock, John, 89, 270, 280, 298; CD's 5 September [1862] letter to, 77, 235
Lyell, Charles, 53, 70, 96, 100, 256, 264; character of, 270–71; CD's 30 July 1837 letter to, 53–54; CD's [14] September [1838] letter to, 54; CD's [2 September 1849] letter to, 87; CD's 4 November [1855] letter to, 81; letter to CD of 1–2 May 1856, 71, 80; CD's 18 [June 1858] letter to, 98; CD's [25 June 1858] letter to, 98–99; CD's 20 September [1859] letter to, 72; CD's [10 December 1859] letter to, 151–

52; CD's 10 April [1860] letter to, 92, 155, 156, 272; CD's 4 May [1860] letter to, 174, 209; CD's 17 June [1860] letter to, 124, 134–35; CD's 3 October [1860] letter to, 120, 124; CD's 14 November [1860] letter to, 168; CD's [1 August 1861] letter to, 136; CD's 12–13 March [1863] letter to, 95; letter to CD of 15 March 1863, 161; letter to CD of 16 January 1865, 285; *Principles of Geology*, 24, 29; and Wallace, 97, 98, 99, 101, 102

Lyell, Mary Horner, 81

Lytton, Edward Bulwer, 84

Mackintosh, James, 271–72

Malay archipelago, 102

Malthus, Thomas Robert, xii, 57, 66, 107, 162; *An Essay on the Principle of Population*, 58

mammals, 193

mandrills, 187. *See also* ape(s)

mankind: as ape *vs.* angel, 161; as architect, 124; arrogance of, 55; baboon as grandfather of, 56; beauty as created for, 128; as created from animals, 55; as descended from apes, 157, 158, 177, 299, 305; as descended from other species, 93; as designer, 133; dignity of, 174; as discoverer, 237; and divine design, 135; and domesticated animals, 83, 122, 125, 138, 186, 211; early progenitors of, 176–77, 193, 206; intellect of (*see* intellect); mental disposition of, 194–95; as mutable production, 173; and natural selection, 182; nature of, 134, 162; origin of as hidden from, 67; origins of, xi, 115, 124, 131, 173–80, 236; population of, xii, 209–10; power over selection, 111, 122, 125, 138, 149, 186 (*see also* animals, domesticated); and preservation of weakness, 211; races of, 181–84, 195; reproduction of, 182; society of (*see* society, human);

mankind (*cont.*)
structure of, 131, 176–77; and struggle for existence, 182, 207–8; and tail, 176

marriage, 59–64, 191–92, 210, 213, 233

Marshall, Victor A.E.G., CD's 7 [September] 1879 letter to, 272

Martens, Conrad, 33

Martineau, Harriet, 281

Marx, Karl, 162, 303

mastodon, 25

materialism, 55

mathematics, 292

matter, origin of, 145

Matthews, Patrick, "Naval Timber & Arboriculture," 92

McCormick, Robert, 37

McDermott, Frederick, CD's 24 November 1880 letter to, 224

medicine, 5, 210, 226. *See also* health

Megatherium, 16

Mengden, N. A. von, CD's 8 April 1879 letter to, 223

Messiah (Handel), 251

metaphysics, 56, 302

mice, 110, 134

migration, of plants and animals, 130–31

Mill, John Stuart, 162

Milton, John, 257; *Paradise Lost*, 34, 254

Mimosa, 289

misletoe, 108

missionaries, 43–44

Mivart, George St. J., 179

mocking-birds, 46

modification, 120, 166; descent with, xiii, 65–72, 92, 129, 131; and domestication, 83; and Galapagos birds, 49; for good of another species, 123; and natural selection, 108; and small *vs.* large areas, 119; and Spencer, 94. *See also* variation/variability

monkeys, 56, 177, 191, 202–3, 258, 272. *See also* apes

Montagu, Ashley, 305

morality, 56, 64, 190–92, 209, 217, 219, 221, 225, 264, 298, 302

Morning Post, obituary of CD in, 299

morphology: and adaptation, 56, 67, 108–9, 111, 142; and animals of Galápagos, 48, 49; and ants, 128–29; and design, 135–36; of embryos, 112–13; of mankind, 131, 176–77; modification of, 83, 131, 166; and natural selection, 124, 142; of orchids, 166; and pigeons, 83; of Primula, 169; as similar in different species, 131

Muller, J.F.T., CD's [before 10 December 1866] letter to, 267

Munchausen, Baron, 16

Murchison, Sir Roderick, 269

Murray, John (publisher), 81, 176; CD's 14 June [1859] letter to, 240; CD's [3 November 1859] letter to, 150

music, 7–8, 223, 257, 291

Napoleon III, 209

nations: competition between, 209; laws, customs and traditions of, 214

natural history, xvi, 14, 15–16, 23, 33–38, 45

naturalists, 11, 12, 15–16, 18, 21, 112, 153, 175, 188–89, 236

natural selection, xi, xiii, 77, 122–26; acceptance of theory of, 96, 175; and adaptive changes of structure, 142; and ants, 129; for benefit of other species, 123; and brain, 174; CD forestalled concerning, 98, 99, 100–101; and CD's reading of Malthus, xii, 58; deification of, 124; and design, 133, 136; and domesticated animals, 53, 111; and extinction, 130; and hive-bees, 199–200; and human architect analogy, 124; and human intellect, 194; as implying choice, 148–49; and individuals of same species, 138; and instincts, 198, 201; as intelligent power, 125–26; and mankind, 182; and Matthews, 92; and modification, 108; and new and improved forms, 130; of orchids, 166; and preservation of varieties, 148; and primogeniture, 233; and progress, 213–14; Shaw on, 302; and structures, 124; and struggle for existence, 98; and suffering, 220; and survival of fittest, 147–49; and United States, 212;

340 INDEX

natural selection (*cont.*)
and variation/variability, 120, 122, 125, 133, 149; and Wallace, xii, 98, 99, 100–101, 102, 147–48, 174, 233

nature, xiii, 115–16, 130, 249–52; adaptation in, 124–25; Conway on, 276; cruel works of, 70; degrees of perfection in, 123; economy of, 56, 70; history of productions of, 114; illimitable schemes & wonders of, 87; personification of, 125, 126; and Sedgwick, 151; selection in, 53; and variation, 120, 122; war in, 57, 66–67; and wedge metaphor, 56, 67, 109. *See also* design; scenery

Nature, obituary of CD in, 300

"Naval Timber & Arboriculture" (Matthews), 92

negroes, 30–32, 187

Nevill, Lady Dorothy, 281

Newman, Henry, 110

New South Wales, Australia, 36–37, 208

Newton, Sir Isaac, 134, 299, 302

New York Times, obituary of CD in, 299

New Zealand, 21, 123

North, Marianne, 281–82

Norton, Charles Eliot, 282

Notebook B (C. Darwin), 54, 55

Notebook C (C. Darwin), 55

Notebook D (C. Darwin), 56, 57–58

Notebook E (C. Darwin), 181

Notebook M (C. Darwin), 56, 256

Notebook N (C. Darwin), 57

observation, xii, xvi, 23, 24, 25, 28, 29, 68, 84, 85, 165, 166, 235, 236, 237, 240, 254–55. *See also* experimentation; science

Ogle, William: CD's 6 March [1868] letter to, 141; CD's 22 February 1882 letter to, 238–39

On the Origin of Species (1859) (C. Darwin), xvi, 107–16, 287–88; barnacles in, 89; CD's plans for, xii; difficulties in, 127, 128–32; Galton on, 277–78; human origins in, 173; Huxley on, 300;

influence of, 298–99; instincts in, 198–200; natural selection in, 122–23; origin of life in, 144; pigeons in, 82–83; and precursors, 95–96; publication of, xi; responses to, 150–64; species in, 117–20, 121; style of, 240, 241; and theism, 222; tone of, 294; too much attributed to natural selection in, 142; and tree simile, 111–12; variation and heredity in, 138–39; and Wallace, 102

On the Origin of Species (1861) (C. Darwin), 92–95, 144–45

On the Origin of Species (1869) (C. Darwin): altered view on selection in, 142

On the Various Contrivances by which British and Foreign Orchids are Fertilised by Insects (C. Darwin), 124–25, 165–67

orangutans, 56–57, 204–5. *See also* apes

orchids, 165–67. *See also* adaptation

organs, 130, 132

Ornithological Notes (C. Darwin), 48

Osorno, volcano of, 27

ostrich, 33–34

Otaheite. *See* Tahiti

Owen, Richard, 26, 155, 157, 272

Pahia, New Zealand, 21

Palaeontological Society, 88

Paley, William, 133; *Evidences of Christianity*, 8–9; *Moral Philosophy*, 8–9; *Natural Theology*, 8–9

Pampas, 19, 69

Panageus cruxmajor, 76

pangenesis, 139, 141, 142, 143, 245

Paradise Lost (Milton), 34, 254

Parana River, 25

Parkes, Samuel, *Chemical Catechism*, 4

Parslow, Joseph (CD's butler), 282–83

Patagonia, 33

Peacock, George, 11

peacocks, 127, 186

Personal Narrative (Humboldt), xii, 10

Peru, 19

Phaedo (Plato), 56

philanthropy, 256
Philosophical Club of the Royal Society of London, 227
Philosophical Society of Cambridge, 22
Philosophical Transactions of the Royal Society, 234
physics, 292
physiologists, 153
physiology, 248
pigeons, 80–83, 140
pineapples, 20
planets, 130, 134, 136
plants, 65, 115, 280; collection of, 165; competition among, 118; and diverse physical conditions, 118; domesticated, 53, 122; genera of, 120; migration of, 130–31; naming of, 165; parasitical, 15; peculiar to Galápagos Archipelago, 165; progenitors of, 144; and struggle for existence, 58; varieties of, 118–19. *See also* botany; population; species
Plato, 95; *Phaedo*, 56
platypus (Ornithorhyncus paradoxus), 36
poetry, 257, 302
politics, 232–33
pollen, 139, 166, 169
Polynesia, 207
poor-laws, 210
population, size of, xii, 57–58, 66–67, 107, 109, 121, 209–10
Port Desire, Patagonia, 33
Portillo pass, 251
Portugal, 84
Portuguese, 32
poverty, 210, 213
Powell, Baden: CD's 18 January [1860] letter to, 91–92; "Essays on the spirit of inductive philosophy, unity of worlds, and the philosophy of creation," 94–95
A Preliminary Discourse on the Study of Natural Philosophy (Herschel), 10
primogeniture, 212, 233
primordial form, 144
primordial soup, 145–46
Primula, 169
Principles of Geology (Lyell), 24, 29
progress, 213–14, 252, 271
proteins, 145
Protococcus nivalis (red snow), 35–36
psychology, 94, 115
Punch, 161

Quadrumana, 176
Quillota, valley of, 26

rabbits, 143
races, 181–84, 205–6, 211
Raverat, Gwen, 296–97
reading, 254, 257, 289, 291–92
reason, 200, 221, 240, 254–55
red clover (*Trifolium pratense*), 110
religion, xiii, 61, 154, 217–25, 271, 280, 298, 303. *See also* God/Creator
The religious views of Charles Darwin (Aveling), 224–25
reproduction, 138, 139–40, 141, 182. *See also* sexual selection
reversion, principle of, 140
Richmond, George, 2
Rio Negro, Northern Patagonia, 25, 33, 34
rock-pigeon (*Columba livia*), 82
Romanes, George J., 301; CD's 7 March 1881 letter to, 197
Rosa (plant), 120
Royer, Clémence-Auguste, 162
Rubus, 120

Ruskin, John, 272
Rydal Water, Lake District, 293

Saint Hilaire, Isidore Geoffroy, 93
Santa Maria, island of, 28
Santiago, Cape Verde, 14
Sarandis River, 25–26
Saturday Review, 24 December 1859, 154
scenery, 15–16, 20, 26–27, 217–18, 222–23, 249–53, 257, 292–93. *See also* nature
science, 62, 71, 158, 232, 234–39, 254, 259, 265; and arguments from results, 68; and brother's chemistry experiments, 4; and causes, 237; and caution, 236; CD's influence on, 299, 300; CD's inspiration concerning, 10; CD's love for, 254, 256; and Chambers, 94; and Darrow, 305; and design, 136; facts, general laws, and conclusions in, 9–10; and false views, 237; and freedom of thought, 224; Gray on, 156; Julian Huxley on, 302; T. H. Huxley on,

science (*cont.*)
153; incomplete and incorrect hypotheses in, 142; and Lyell, 271; nonbiological, 292; Sedgwick's encouragement in, 22. *See also* hypothesis; observation
scientific laws, 10, 54, 93, 114, 115, 116, 126, 133, 145, 303; of battle, 185; and CD's mind, 253, 258; designed, 134, 135; general, 124; and human origins, 173; secondary, 166; and variability, 128. *See also* gravity
Scott, Sir Walter, 291–92
séances, 238
sea-sickness, 14
Sebright, Sir John, 83
Sedgwick, Rev. Adam, 9, 22; letter to Darwin, 24 November 1859, 151
seeds, 108–9, 139, 141
sentient beings, 135–36, 219
sexes, 185–88. *See also* gender
sexual selection, 181, 183–84, 185–89. *See also* reproduction
Shakespeare, William, 257

Shaw, George Bernard, *Back to Methuselah*, 302
Shelley, Percy Bysshe, 257
shells, 65
shooting, 6, 7
Shrewsbury, gravel beds in, 9
Shrewsbury School, 3, 4, 6
Silas Marner (Eliot), 291
Silk, George, 160
slavery, 30–32, 61, 232, 233, 264
snuff, 281, 290–91
society, human, 207–14, 233
soul, 56, 222
South America, xi, 11, 26, 28–29, 69, 107, 234
species, 53–58, 68, 71, 117–21, 235; adaptation by (*see* adaptation); classification of, 65, 70, 89, 112; and community of descent, 117; competition between, 111, 119; as created, 71, 92, 114, 285; definition of, 67, 117, 118; descendants of, 108; descent of, 93, 95; difficulty of distinguishing, 94, 117–18; diversification of, 119–20; domesticated, 53, 122, 125, 138,

139, 140; dominant, 119; exactness of, 55; extinct, 26, 84, 108, 110, 111, 114, 129–30; increase of, 66–67, 120; and individual creations, 67; individuals of same, 138; modification of (*see* modification); as mutable productions, 173; new, 58, 95, 97, 119, 129, 136, 149; and new conditions of life, 120; as not immutable, 65; origin of, 57, 93, 102, 107, 114, 155, 285; and Powell, 95; relation of oldest forms to existing, 150; and selection in nature, 53; and Spencer, 94; stability of, 48; and sterility, 67; and struggle for existence (*see* struggle, for existence); and sub-species, 117; survival of (*see* survival, of fittest); theory of, 85, 86; transmutation of, 54, 69; and tree simile, 111–12; variation among (*see* variation/variability); warring of, 57–58; weaker structures in, 56. *See also* animals; plants; population

S. Pedro, Chonos Archipelago, 35
Spencer, Herbert, 147–48, 272–73; "Essays," 94
spiritualists, 237–38
Sprengel, C. K., *Das entdeckte Geheimnis der Natur*, 57
Stazione Zoologica di Napoli, 238
St. Jago [Santiago], Cape Verde islands, 24
struggle, for existence, xii, 107–8; and American Civil War, 233; and Malthus, 57–58; and mankind, 182; and natural selection, 98, 123, 148, 149; and population size, 66–67, 109, 110; and races of men, 181
sublimity, sense of, 222–23
subsistence, 209–10
suffering, 219–20, 233. *See also* vivisection
Sullivan, Bartholomew James, 225
Sundew, 289
survival, of fittest, 142, 147–49
Sutton, Seth, 203

Tahiti, 20, 43, 291
taxidermy, 5

theism, 222, 224, 265, 271
theology, 303. *See also* religion
Thompson, Sir William, 163
Thorley, Catherine A., 165
Tierra del Fuego, 39–42, 44, 178, 217, 250
The Times, 152, 232; obituary of CD in, 298–99
Timiriazev, Kliment, 283
toads, 34
Tollet, Ellen, 225
Tollet, Laura, 225
tortoises, 37–38, 45–46
Toxodon, 26
Tree of life, 112
Turin Society, 238
Twain, Mark, 284
Tyndall, John, 284
tyrant-flycatchers, 46

Unitarianism, 265
United States, 158, 209–10, 212; Civil War in, 232, 233
universe, 135

vaccination, 210–11
Valdivia, Chile, 27–28
The Variation of Animals and Plants under Domestication (C. Darwin), 83, 125–26, 136–37, 139–41, 142, 148–49
variation/variability, 66, 138–43, 150; and adaptation, 67; and chance, 139; and conditions of life, 128; and design *vs.* natural selection, 133; and human selection, 122; as independent of conditions of life, 121; and instincts, 201; and laws, 128; and natural selection, 120, 122, 125, 133, 148, 149; and organic and inorganic conditions of life, 122; and pigeons, 82–83; slight degrees of, 88; and struggle for existence, 58, 67, 107–8; and suffering, 220; and use and disuse, 128; useless, 142. *See also* modification
variety (taxonomic), 117, 118, 119
Vestiges of the Natural History of Creation (Chambers), 91, 93–94, 154
Victoria, Princess Royal of Russia, 285
vivisection, 246, 247–48. *See also* suffering
Volute (sea snail), 9

Wallace, Alfred Russel, 95, 116, 304; admiration of for CD, 156, 160; CD as forestalled by, 98, 99, 100–101; CD on modesty of, 157, 273; and CD on sexual selection, 181, 182; CD's 22 December 1857 letter to, 97, 173, 235; CD's 18 May 1860 letter to, 157; CD's 28 [May 1864] letter to, 181, 233; CD's 15 June [1864] letter to, 182; CD's 5 July [1866] letter to, 147–48; CD's 27 March [1869] letter to, 142, 175; CD's 14 April 1869 letter to, 175; CD's 20 April [1870] letter to, 273; CD's agreement with, 97; on completeness of CD's work, 156; on Darwin's abilities, 284–85; letter to CD, 2 July 1866, 125, 147; letter to J. D. Hooker, 6 October 1858, 101; letter to Charles Kingsley, 7 May 1869, 284–85; letter to George Silk, 1 September 1860, 160; and natural selection, xii, 98, 99, 100–101, 102, 147–48, 174, 233; on similarities to CD, 301–2; *On the Tendency of Species to Form Varieties; and on the Perpetuation of Varieties and Species by Natural Means of Selection*, 100; simultaneous announcement of theories with CD's, xii, 99, 100, 101, 102
Waterton, Charles, 5
Watson, James D., 305–6
wealth, 211–12
Wedgwood, Caroline Sarah Darwin (aunt), 225
Wedgwood, Hensleigh, 238
Wedgwood, Josiah, II, 11–12, 22, 273–74
Wedgwood, Julia, 287; CD's 1 July [1861] letter to, 135–36
Wedgwood family, 296–97
Whewell, William, 67, 142, 160
Whitley, C. T., CD's 8 May 1838 letter to, 59
Wickham, John Clements, 13, 30
Wilberforce, Rev. Samuel, 157, 158
will, 137, 204

Wilson, William, 43
Wollaston, Thomas V., 71
wolves, 65
woodpeckers, 108, 165, 252
Wordsworth, William, 257; *The Excursion*, 254. *See also* Lake District, UK
worms, 115
writing, xv, xvi, 14–15, 240–43, 289, 290, 293–94

Yaghan people, 41
York Minster (El'leparu), 42

Zoological Gardens, London, 203, 204
Zoological Society of London, 35
zoology, 4, 45–49, 91
Zoonomia (E. Darwin), 90, 93

More Princeton University Press
QUOTABLES

Available wherever books are sold.
For more information visit us at www.press.princeton.edu

www.ingramcontent.com/pod-product-compliance
Lightning Source LLC
Jackson TN
JSHW021708070225
78575JS00011B/6